The Thermodynamics of Pizza

The
Thermodynamics
of Pizza

Harold J. Morowitz

Rutgers University Press
NEW BRUNSWICK AND LONDON

These essays have been previously published, some in slightly
different form, in *Hospital Practice.*

"The Cobra," by Ogden Nash, from *Verses,* copyright © 1931 by
Ogden Nash, is reprinted by permission of Little, Brown and
Company.

Library of Congress Cataloging-in-Publication Data
Morowitz, Harold J.
    The thermodynamics of pizza / Harold J. Morowitz.
        p.   cm.
    ISBN 0-8135-1635-8
    1. Biophysics. 2. Science. I. Title.
QH505.M62 1991
500—dc20   90-8674
                                                               CIP

British Cataloging-in-Publication information available

Book designed by Liz Schweber Doles

Contents

Metaphysical Musings

Fauna

Societies

The Computer Revolution

The Thermodynamics of Pizza

Thought for Food

The Thermodynamics of Pizza

We were chatting at lunch one day, when one of my companions complained of burns on the roof of his mouth from the previous evening's pizza party. Talk drifted in two directions: one dealing with the sociology of pizza as the all-American dish and the second focusing on why pizza stays so hot for so long. The first topic didn't seize my attention, but the second suggested a largely unexplored scientific domain, the thermal physics of pizza. To establish preeminence in that field I went to the library, then to the computer, and finally to the family cookbooks. My first brief contribution was hurriedly submitted to Quick Publications in Culinary Physics and is reprinted below.

To a first approximation an uncooked pizza consists of an array of three cylindrical disks: A, B, and C. Disk A is made of a yeast-flour dough and is of thickness α. Disk B is made largely of tomato paste and is of thickness β. Disk C is made of mozzarella cheese and is of thickness γ. In general $\alpha > \gamma > \beta$. Trace materials, such as oregano and pepper, are also present in small quantities and will be considered in our preliminary treatment only with respect to surface effects. The uncooked trilamellar pizza is rapidly placed in an effective infinite isothermal reservoir at 533 °K (Kelvin absolute thermodynamic temperature).

Three changes occur while the stacked disk structure equilibrates with the high-temperature oven. First, the layer of dough bakes into bread, a low-water-content material with a large number of nonconnecting small air spaces. Second, the tomato paste dehydrates, and third, the mozzarella undergoes a complex series of transitions involving protein denaturation and lipid rearrangement from regular liquid crystals to more disordered states. Those transitions undoubtedly contribute to the high heat capacity of mozzarella.

Some background information is necessary to understand the structure of mozzarella. The manufacture of that cheese begins with an initial fermentation process, as is the case with all other such products. However, when the necessary acidity has been reached, the curds are separated from the whey and made plastic by heating and kneading in hot water until the material can be stretched in ropelike strands or threads. The pre-cheese at this stage is known as *pasta filata*, a generic term that also applies in the manufacture of provolone. Clearly, a more thorough understanding of our problem will eventually require x-ray fiber diffraction of *pasta filata*, but that study will have to await a stronger interest and greater funding for gastronomic physics.

After removal from the oven, the pizza is sliced and placed quickly in a flat cardboard box, which is immediately closed and often taped shut. There is no physical separation after the slicing, so that edge effects can be ignored and we can treat the pizza, for thermal purposes, as an infinite plane. The procedure reduces the heat-transfer problem to one dimension represented by a vector normal to the pizza surface.

The bottom layer is now a piece of cardboard, which is a very poor conductor of heat. The next lamella is the baked pizza dough, which is an excellent insulator because of the unconnected gas spaces throughout. With respect to the thermal properties of breadlike materials, note how toast burns on the outside with little evidence of damage below the most external portion. As a point of historic interest, the *Dizionario Etimologica Italiano* (Firenze, 1954) traces the Italian word "pizza" to the Greek πιττα, a bread that is also achieving considerable importance in the American diet. The flow of culture from Greece to Rome shows up in the most unexpected ways.

Regarding the thermal properties of partially dessicated seasoned tomato paste, the literature is most unhelpful. It is surprising in this age of science that there is still such a paucity of available data in many areas of technologic importance. We are thus reduced to guesswork in engineering domains in which full theory exists to account for the empirical information. From compositional considerations the tomato layer would be expected to have a relatively high heat capacity and low conductance. It thus serves as a buffer between the mozzarella and the baked dough.

The melted mozzarella layer, which we shall designate MML, is the obvious source of trauma to the roof of the mouth and from the point of view of medical physics is clearly the key agent in the etiology of pizza burn. On the bottom side, the MML is very well insulated by the bread and cardboard layers. To a first approximation all heat loss will be upward from the surface to the covering cardboard. Note that the intervening material, a thin layer of air, is also well known for its insulating properties.

The loss of energy from the upper surface of the MML occurs by three processes: radiation, convection, and conduction. Radiation could be precisely determined, in principle, if we knew the reflectance of mozzarella and cardboard in the long-wavelength regions of interest. The process is clearly governed by the Stefan-Boltzmann fourth power radiation law. Here, the oregano and other trace materials sitting on the surface might be of some importance if they changed the surface radiance in any appreciable way. In any case, the temperature range is such that radiation is likely to be of secondary importance. Convection also plays a minor role because of the relative thinness of the air layer and the presence of the upper cardboard heat shield. The insulating properties of the cardboard play a major role, since they serve to minimize the temperature gradient across the airspace. By studying the cooling of pizza in the gravity-free environment of the space shuttle, we could more exactly determine the importance of convective flows.

The crucial parameter in the entire problem—the thermal transfer coefficient between melted mozzarella and air—is also subject to considerable uncertainty. The absence of even rudimentary numerical values of the key factors makes it impossible to attempt serious mathematical modeling of this problem. We would probably do better at this point to work back from the cooling time of pizza to an estimate of the critical parametric values. Nevertheless, it seems most likely that simple thermal conduction across the airspace is the major process in the cooling of boxed pizza. We can qualitatively understand that the MML retains its burning abilities for a long time for two principal reasons: 1) MML starts out at a very high temperature. The pizza oven is a full 160 °C higher in temperature than boiling water. 2) The mozzarella layer is embedded between insulating lamellae and thus loses heat very slowly.

The problem has, of course, yielded to scientific analysis and seems in no way to involve a change in paradigm.

Beyond the thermodynamic view is the gastronomic one. What is the optimal temperature of pizza for eating pleasure? Here the necessary experiments are clear-cut, and I for one would happily volunteer to be a member of the taste panel if it were agreed to start from the lower temperatures and work up. After all, who wants to try pizza at 533°K?

Circles
of Reality

E very so often one tires of the sophistication of modern
day-to-day life and longs for simpler pursuits: respites
from the sin of taking oneself too seriously. I was in just
such a mood when a visit from our eight-year-old grandson
gave me the opportunity to seek a view of reality through
the eyes of a child. In deciding what we were going to do,
we somehow—and I can't tell exactly how—decided to
bake bagels.

It's also not clear why we chose bagels rather than bread
or brownies, but I suspect Matt and I had different motives.
He likes bagels; they're one of his favorite foods. I also like
them, but I suspect I was influenced by the thought that
my great-grandfather, whom I never knew, was a bagel
baker. An activity spanning six generations from my great-
grandfather to my grandson had a certain romantic appeal.
In these days of familial transience, whenever you can get
your hands on six-generation stuff you'd better grab it and
knead it for all it's worth.

Bagel baking in Kapulya, a small town in western Russia,
could not have been the most lucrative of professions, for
I have often heard great-grandfather referred to as a poor
man. His daughter had married well—to a successful carpenter and landowner—and he liked to visit her. He often
remarked that since his home was so small he hoped to die
in her house so there would be adequate room for memorial
services. And sure enough, one day while visiting he came
down with pneumonia. He was put to bed in his daughter's
house and died two days later. Everyone wondered how
he had managed that feat, but it was too late; there was
no way to question him.

In any case, two totally inexperienced bakers set out with only a recipe between them and total ignorance. The old family formula has, alas, been lost. The ingredients— vegetable oil, flour, sugar, salt, egg, water, and yeast— seemed simple enough. We began.

To an old microbiologist, the yeast was the most interesting part of the mixture. Adding the leaven reminded me of a visitor who came to our lab to spend a sabbatical. Although he had traveled halfway around the world, he carried with him a yeast culture—not for his experiments but to continue his ritual of baking bread once a week. How we relished invitations to his parties, where hors d'oeuvres of bread and butter or bread and jam were the gustatory high point! He would frequently remind me that the best of food and drink involved fermentation, citing bread, cheese, and wine. In a sense "A Jug of Wine, a Loaf of Bread—and Thou" in one way or another involve yeast in all three parts.

Yeasts are most unusual organisms, being the simplest of all eukaryotes. At some early evolutionary stage a split occurred: One evolutionary branch led to modern bacteria with their great facility in biochemical energy processing; the second branch led to meiosis and sexuality. The yeasts, or close relatives, were the first step on the road to sex.

All of this yeast lore made little difference as we got our hands into the mixture and found it was much too sticky to knead. Indeed, it stuck to the hands like the old flour- and-water paste I remembered from childhood. Luckily, I also remembered reading papers on the adsorption iso- therms of water and flour. Deducing that the problem was too much free and too little bound water, we added flour and marveled as the mixture was transformed into dough. Kneading it then was one of those mild sensual pleasures that we both thoroughly enjoyed.

We then formed the dough into circles and set it aside

to rise. In those moments of leisure I thought about a phone call I had received many years ago from the university chaplain.

"Say, Harold, I'm trying to convince our housekeeper, who is a teetotaler, that alcohol is produced when bread rises, and she won't believe me. Would you speak to her?"

Before I had a chance to reply, the lady in question was on the phone and I had to break the news to her that a small amount of ethanol is indeed produced in the rising of bread. I quickly assured her that the heat of baking drives off any alcohol, so that the final product is free of the chemical she so opposed. She thanked me and the chaplain thanked me.

When the bagels had risen we proceeded to the next stage—dropping them into boiling water. The problem with bagel recipes is that they are somewhat hazy about the boiling stage, particularly as to how long the objects should be so treated. We guessed our way through that one and went on to the next stage—twenty-five minutes in a 400° oven.

Then out they came, hot and delicious. I was excited at our modest achievement. Matt, in his childish honesty, told me they tasted more like rolls than bagels. I assumed we hadn't boiled them long enough. We will have to go back and make more batches in an effort to achieve the perfect bagel. But no matter, we have spanned six generations, and that is more important than the exact nature of the products.

I'm left trying to make contact with what went on in the mind of my great-grandfather as he mixed the dough, watched it rise, boiled the circular pieces, and then placed them in the oven. The world has changed so rapidly, it is hard to break the time barrier between me and someone I never knew. The best of historical writers have been well aware of the difficulty of understanding the mind-set of another age.

I can, however, recommend the therapeutic value of mixing flour, water, yeast, and whatever else one uses. Leavened wheat has been so much a part of human culture for so long that this exercise connects us with our forebears over hundreds or thousands of generations. Occasionally, that kind of reaching back into the past is a good experience. At least Matt and I enjoyed it.

The Olive
in the Martini

I find myself at a cocktail party looking around for a familiar face. None appears, but a kindly lady with a tray of martinis comes on the scene, and I cannot be so rude as to refuse her offering. In my solitude, I stare at the glass before me and begin to muse over the technological triumph represented by the contents of that vessel.

Fermentation appears to be as old as civilization. All cultures known to anthropological science have developed some form of alcoholic beverage. Mesopotamian pottery from 4200 B.C. depicts brewing scenes. Wine is mentioned no less than 186 times in the Old Testament. (How's that for a piece of cocktail party knowledge?) By A.D. 800, the monks of Europe were vintners of considerable skill and productivity.

The early history of distillation is as obscure as that of fermentation. There is some learned speculation that "cloth and sheepskins were held over boiling fermented mash to condense the alcohol, which was then wrung out of them." The current techniques moved from the early Egyptian alchemists to the Arabs and then into Europe in about the twelfth century. The first written description is by Master Salernus, who died in A.D. 1167 (Dark Ages indeed). Thoughts of evaporation and condensation trigger my thirst, and I take a sip, silently toasting Arnold de Villeneuve, the unsung savant who wrote the first full treatise on distillation somewhere around A.D. 1300.

The early spirits known as aqua vitae (the water of life) were scarce and expensive medicines. Three centuries passed from de Villeneuve's classic work to the inventive triumph of Francis de la Böe, professor of medicine at

Leyden University. This seventeenth-century sage added oil of juniper berries to aqua vitae, producing a drink he named *genièvre* (*juniper* in French). This became *geneva* in Dutch and was finally shortened to the English *gin*. Modern varieties are alcohol-water mixtures redistilled over juniper berries, to which may be added other botanicals, such as angelica root, anise, coriander, caraway seeds, calamus, cardamom, cassia bark, and orrisroot. A very substantial plant taxonomy was necessary to produce the present product, and as may be expected, the exact formulations are carefully guarded secrets.

Although gin is the chief ingredient of martinis, a second alcoholic beverage is added: the wine vermouth. Indeed, the name of the cocktail comes from the noted vermouth manufacturer Martini & Rossi. There are no fixed standards as to the volume ratio of vermouth to gin. The dryness of a martini depends on the smallness of this quantity. For those devoted to quantitative measure, we may take the negative of the logarithm of the vermouth : gin ratio to define dryness. A very, very dry martini contains less than one part in 10^5; this may also be referred to as a homeopathic martini. A mystically dry martini is made by passing a sealed bottle of vermouth by a sealed bottle of gin. A very sweet martini can contain up to 33 1/3 percent vermouth. A taste of the drink before me indicates that this establishment has a very small wine bill.

If gin is botanically complex, vermouth involves a virtual compendium of plant taxonomy. As many as sixty herbs go into the secret formulations. One starts with fortified wines (such as sauterne), mixes them with herbs, extracts the plant material, and then separates off the clear fluid. Prominent among the herbs is wormwood (*Artemisia absinthium*). The very name *vermouth* probably derives from *Wermut*, the German word for *wormwood*. As I recall that this substance in high concentration can cause "delirium and hallucinations," I am just as happy that my host's bartender has opted to make his martinis decidedly on the dry side. After all, one still hears horror stories of Henri Pernod's absinthe, which was very rich in extract of *A. absinthium* and led to very many "bad trips."

Of course, the liquid is only part of martini technology. Resting quietly in the bottom of my glass is a large green olive, a fine specimen of the fruit of *Olea europaea*. I have long wondered how the olive became edible. The present method of processing green olives illustrates the problem. The fresh fruit contains a glycoside so bitter as to make it inedible. Unripe green olives are harvested and immediately placed in a dilute solution of lye and allowed to soak. The hydroxide is then removed by several washings, and the fruit is pickled in a strong salt solution. Sugar is added after several weeks to maintain the fermentation. After six months, the olives are bottled or canned.

The question that comes to mind is how anyone ever discovered treatment with lye to get an edible product. The answer is lost in antiquity, for before Homer, before the Bible, in the dim reaches of the past, olives were grown, treated, and eaten. There is evidence of the cultivation of olive trees on Crete going back to 3500 B.C.

Gazing down at my glass, I see a flash of red color that reminds me of the latest history of the olive: the coring machine and the stuffing machine. The pimento (*Capsicum annum*) at the center also has a long story of cultivation and preparation. This knowledge, alas, has also been lost to culinary history.

The feel of the glass in my hand suggests the final technological development necessary for the modern martini: adequate home refrigeration. I can remember in my early childhood the vanishing iceman still making some neighborhood deliveries. The modern compression-expansion refrigerator had its beginning in the thermodynamics and development of electrical motors of the late 1800s. By the first third of the twentieth century, cold was readily available and purists could place gin in the freezer, avoiding even the contamination of small amounts of pure water.

I look up. Someone is approaching me and wishes to engage in conversation. I begin, "Say, my friend, have you ever thought of the technological triumph of the modern martini?"

Ecology and the Great Cold Cereal Rip-off

A supermarket may not seem like the ideal setting for the study of ecology, but that is only because of a regrettable tendency we have to separate our academic thoughts from the nitty-gritty of everyday life. If I remind you that I am here to purchase chemical-free energy for oxidation-reduction reactions, fixed nitrogen, necessary minerals, and biochemicals that I am unable to synthesize, then my travels down the aisles would more closely resemble a field expedition to some exotic environment. However, having placed the matter in a scientific context, I must continue to look for meaning in every aspect of the situation—including the unit prices that are flashing past my eyes.

Since the 1940s ecologists have recognized that the prime medium of exchange in most ecosystems is energy. And so they have classified organisms into producers, those species that directly take the sun's energy and store it in a chemical form largely consisting of sugar and carbohydrate; herbivores, who live by eating producers or parts thereof; and carnivores, who survive by eating herbivores. There are then top carnivores and the like, but they are of little interest in grocery-store ecology.

Now, for reasons of biological efficiency it usually takes several pounds of producers—plants—to make a pound of herbivore. Thus, we need several pounds of grain to make a pound of cow or chicken. Therefore, by the inexorable laws of economics, meat should be more expensive than grain. Carnivores by the same reasoning should be more expensive, but we rarely eat carnivore flesh except when we catch it wild—as by the fishing industry. If we were to farm foxes for meat and feed them cows, then the price of fox steaks would be very high indeed. We do farm carnivores for fur, but that is well known to be an expensive commodity.

Getting back to the still empty cart before me, I should expect a certain uniformity of price in plant materials unless they are rare or difficult to grow or harvest, such as certain spices. Of course, my analysis is complicated because foods have differing water contents, so unit prices should ideally be based on dry weight. But that's a lot of science to expect out of a grocery store.

In any case, I would like to concentrate on primary-producer carbohydrate prices, particularly grains and sugar, as those constitute the major components of most diets. I'll talk mostly about foods that are sold dry so as not to confuse the issue with concern about differing water content. Sugar is now selling at 38 cents per pound. Wheat flour is between 17.6 and 34.5 cents per pound, depending on the brand and the way it is hulled. Rice is between 33 and 87 cents a pound, depending on brand and on processing. Cornmeal has two brands at 33 and 36 cents per pound, and barley ranges from 39 to 45 cents per pound. There is a fairly surprising uniformity of price of this first trophic level grain material, with a mean of about 35 cents per pound.

In my ecological economics mood I still haven't managed to put anything in the pushcart, so I wander down to the meat section, checking prices and making a guess about the exact water content of the various cuts of meat. Meat costs between $3.00 and $4.00 a pound dry weight. That's about ten times the cost of grains. That factor is most interesting; if I recall the ecological literature, the efficiency of conversion from producer weight to herbivore weight is about 10 percent, so that farm-raised meat should be about ten times as expensive as the grain eaten by the animals. Thus far, our theory is working remarkably well. The supermarket is right in line with biological theory.

But economics often goes beyond straight biology, and I have another question. What happens to the primary producer material when it is processed rather than being converted to meat? Here we note that bread ranges from 50 cents to a dollar per pound; pasta is 63 to 95 cents per pound. Oatmeal is 76 cents and Cream of Wheat is 97 cents per pound. We note a mean value of about 70 cents a pound for prepared primary-producer material, or a factor of about two over the starting material, which sounds reasonable enough.

The logic of my ecological economics would have been just fine if I hadn't turned into the cold cereal aisle. Here I found some real surprises.

Cereal	Price per pound
Corn Flakes	$1.40
Raisin Bran	1.59
Fruit Loops	2.34
Puffed Rice	2.49
Mr. T	2.68
Cocoa Puffs	2.80
Puffed Wheat	2.91
Trix	2.98
Total-Corn	3.18
Request Pack	3.81

This material was averaging about $2.10 per pound, or three times as much as one would expect on the basis of its agricultural type, ecological level, and state of preparation. A factor of three in the price of a major American breakfast food is a matter of some importance to the economy.

Where does this anomaly come from? A clue is found in a trip to the generic brand section, where the puffed wheat that sold at $2.91 with a brand name on the box is going at $1.20 per pound. In other words, if I buy the name brand I am paying more for the name, the printing on the box, and the advertising than I am paying for the contents. Even with that vast discrepancy, I suspect that the generic puffed wheat is overpriced because people are used to such inflated prices in cold breakfast cereals.

In the end, I must advance the hypothesis that something is rotten in the cold cereal industry. Large boxes full of very little are commanding extraordinary prices, in part, I believe, because people haven't stopped to think about what cold cereal really is.

In any case, a food substance that is overpriced by a factor of three should be looked at with grave suspicion by the grocery buyer trying to get the best nutrition for the available money. Think of all those people on food stamps wasting their grocery dollars on so little value. Perhaps we should teach grocery-store ecology in the schools.

By now I've lost my appetite. The cart is still empty. I go back to dried peas and beans at 45 cents a pound and up. That has got to be the best nutritional bargain in the store. In my present mood, such a deal is irresistible. I leave with twelve pounds of those dehydrated legumes, wondering how we are going to cook them.

Coffee

About once every three or four years I spend a few idle moments at the library checking out "coffee." I try to keep up with the medical implications of this substance just to assure myself that there is nothing really wrong with my habit. One likes a little periodic reassurance that one's actions are not more irrational than is absolutely necessary.

My interest in the physiology of this common drug dates back many years to an overnight hike taken with my number-one and number-three sons. We had walked through Haleakala Crater on the island of Maui and had slept in sleeping bags, and we were stopping for breakfast during our climb out. I had with me a plastic envelope of instant coffee and we, of course, carried a canteen of water. I had already resigned myself to a tepid brew, as fires were not permitted in the national park. However, the sudden realization hit that we had nothing to mix the coffee in. My hiking companions took this news very lightly, but I realized that for me it was a serious matter.

We sat there breathing the wonderfully clear air at 10,000 feet and chewing on our dried-fruit breakfast. In spite of the absolute tranquillity of the surroundings, my mind raced as I tried to resolve the coffee problem. Suddenly I reached into the knapsack and withdrew the Frisbee we had brought for recreation. Coffee was measured out into the flat vessel, water was poured in from the canteen, and I sipped and savored the muddy solution to the roaring laughter of my unsympathetic children. It was at that point in space and time that I realized that coffee is an addictive drug. Most users never face up to the habit-forming nature of the substance because they never try to withdraw. For some years afterward, I would occasionally be greeted at breakfast by the query, "Have you had your fix yet today, Dad?" Such are the trials of parenthood.

However, addictive does not necessarily mean deleterious, except in the Thoreauvian sense that we sacrifice our freedom to those things that we are unable to do without. Nevertheless, reports appear from time to time suggesting that coffee may not be as benign as it seems to be when we savor the wonderful odor of the roasted beans. Thus, I periodically return to the literature.

My efforts to find the information I wanted in *Index Medicus* proved interesting in themselves. First, coffee is indexed alphabetically between *coercion* and *cognition*. That is an accident of language, of course, but it does have a poetic ring of reality about it. Less accidental are the subheadings under coffee: Adverse Effects, Analysis, Immunology, Poisoning, and Toxicity. It's clear from the language that the editors of *Index Medicus* have taken a pretty firm stand on whether or not the stuff is harmful.

The editors of *Lancet* are less doctrinaire. In the January 31, 1981, issue they noted that "there is no evidence that excessive drinking of coffee causes any physical harm." They even acknowledged that "there is some evidence that caffeine protects against cancer." Of course, there are also lots of psychological effects resembling anxiety neuroses, and the myriad reports on the withdrawal syndrome give experimental confirmation to my Frisbee-slurping behavior. Ten or more cups a day are reported to cause nervousness, irritability, agitation, headache, and muscle twitching.

Research has moved apace, and we now know that caffeine acts by inhibiting the enzyme cyclic-nucleotide-phosphodiesterase, thus enhancing the effect of the second messenger, cyclic AMP. In the January 20, 1983, issue of *Nature* we learn the even more startling news that both normal and decaffeinated coffees contain a substance that blocks opiate receptor sites in the brain. Coffee is thus getting into the physiologic system in rather subtle ways. The psychopharmacology of our most popular drink is clearly in its infancy.

All of this thought about opiates set me to thinking about the history of the human use of coffee, which seems to be a most unlikely commodity to rank second after oil in international trade. The present methods of processing and consuming this material could hardly have come about quickly.

Legend has it that a certain shepherd of Arabia was impressed when he noted the animated behavior of his sheep who were chewing on the berries of a certain shrub. He began chewing on it himself, and mankind has been hooked ever since. That legend is perhaps as close as we will ever come to the early history of our subject.

The second stage in coffee technology was getting rid of the berry and concentrating on the seed, or bean, which contains about 2 percent caffeine by weight. Roasting was quite likely invented by an accidental fire involving coffee berries. The delightful aroma of the heated seeds could hardly have passed unnoticed. Grinding the beans and making a hot aqueous extract, while less obvious, does have other precedents. By the 1700s coffee had spread from the Near East to Europe, and the rest is history. Coffeehouses and culture and coffeehouses and conspiracy are recurring themes from Istanbul to Berkeley.

Just when I was about to totally relax about my habit, since the literature had so little bad news, I came across an article in *Mutation Research* entitled "Roasting coffee beans produces compounds that induce prophage X in *E. coli* and are mutagenic in *E. coli* and *Salmonella typhimurium.*" In those very interesting experiments by a group at the Cancer Center in Tokyo, controls were carried out with coffee prepared from green beans. Since that brew lacked prophage-inducing activity, the researchers conclude that lysogenic and mutagenic compounds were produced in the roasting process.

I'm now being driven back to my "moderation in all

things" attitude. It's clear that a commodity that is number two in world trade can hardly be an acute toxin. It's also clear that I have little desire to extract green coffee beans and drink them. My tentative conclusions and advice after this brief literature review are, therefore:

- Keep your consumption far below ten cups a day.
- Check out the literature in another three or four years.
- Never, I repeat, never feed roasted coffee to your bacterial cultures.

Feast, Famine, and Fetish

O ur friend Virginia was reporting to us on her new job of transforming carrots and bell peppers from the solid to the liquid state. Her employment in a "health food" emporium gave her an opportunity to observe a novel slice of life, and she was converting her macerating task into a learning experience. She noted that the clientele were often wan, emaciated, and obsessively concerned with the condition of their bodies. She could not help comparing them with the jovial, weightier customers at the pizzeria across the street.

Virginia's story brought to mind a previous visit to a college campus where I had found one of our son's friends systematically starving himself to avoid ingesting sugar. I delivered a two-hour lecture on sucrose, glucose, fructose, and the glycolytic cycle. Mitch gradually became convinced that the molecules he was avoiding as dimers were indeed contained in high-polymer form in the food he was eating. In the body these substances are broken down to yield the same glucose and fructose that are released from the digestion of cane sugar. The young man recanted, and we celebrated by stuffing him with hamburgers and extra thick milkshakes at a fast food outlet. It was a joyous feast.

I mentally filed away the Mitch and Virginia sagas, for I have often expressed a biochemist's curiosity about which foods are healthy and a skeptic's doubts about how we know which foods are beneficial. The issue came to the fore a few months later when we had as a houseguest a professional scientist who approached most problems as a hard-nosed realist. He was, however, so devoted to boiled grains that I could not resist the thought that our omnivores' bill of fare must have seemed somewhat distasteful to him. The kitchen staff (my wife and I) tried to respond, but it was a bit of a burden to those of us fully employed at other jobs.

My next thought on this matter occurred while I was a houseguest of a dear relative who followed a doctrine that proclaimed the noxious quality of hydrocarbon (fat, oil, wax) in animal, mineral, or vegetable form. I have had a lot of experience with polymers of CH_2 groups and find it difficult to view them with alarm, but clearly I was again being treated to a very important Truth-in-Eating experience.

I know that we are all influenced in our thoughts about nutrition by events out of our dietary pasts, so I had best confess my prejudices at the outset. The healthiest individual I have known was my paternal grandmother, who lived to age ninety-six. I often lunched with her on Saturdays, and the meal frequently began with a slice of raw turnip liberally smeared with chicken fat. Just the memory of it evokes a Pavlovian flow of saliva and lipases. The chicken soup, which followed, always had a thick oil slick floating on top. By the rules of many contemporary nutritionists her diet was unadulterated poison; yet the toxic effects took ninety-five years to manifest themselves. This is the kind of childhood experience that, while statistically insignificant, does tend to dull one's dogmatism.

The second major influence on my views is less anecdotal and much more satisfying to statisticians. The surrounding events occurred nineteen years, four months, and three days ago when I was giving up smoking and was disciplining myself by regularly reading and rereading the celebrated American Cancer Society report: "Smoking in Relation to Mortality and Morbidity" by Cuyler E. Hammond (*J Natl Cancer Inst* 32:1161, 1964). In the process of convincing myself of the evils of cigarettes, I came across some cross-correlations that were quite surprising. For example, the data revealed that, for both smokers and non-smokers, the mortality rate decreased with the number of meals per week of fried food. In other words, if longevity is your goal, eating fried food is good for you. This was a most counter-intuitive result, yet it came from the most thorough statistical study ever carried out on life-style and health, one involving 400,000 men over a three-year period. I have tried to interest nutritionists and internists in this result for the past nineteen years. This has been done in conversation, in public lectures, and in print. The sole response has been a reprint request from a fried chicken chain.

When social conversation turns to food and health and I mention my grandmother or Dr. Hammond, I find that mild-mannered people become abusive, quiet individuals produce profanities, and peace advocates have to be restrained from acts of violence. The types of reaction evoked by discussion of food and health have an intensity and emotional depth characteristic of conversations about religious belief and, in truth, there is a certain philosophical nexus.

It has long been established that religious and scientific certainty come in very different ways. Religious belief comes from acts of faith, an approach that is not open to public verification. Scientific knowledge comes from a network of experimental results and theoretical constructs that are open to examination, and repetition or failure to repeat, by all interested parties. If nutritional dicta are derived from faith, the rules are then in the nature of dietary laws and should be treated with all the respect one owes to another's theological opinions. If nutritional rules come from science, then they are subject to experiment and critical attempts at falsification, which is the acid test of all natural philosophy.

Before launching into a discussion of epistemological ideas about how nutritional information is established, I want to cite two recent examples from the literature indicating the uncertainty of recent medical advice on nutrition. The first is a news article in *Science* ("Dietary Dogma Disproved," April 29, 1983), which discusses the work of Phyllis Crapo in establishing the fallibility of the dictum: "Complex carbohydrates, such as the starches found in rice and potatoes, take longer to be absorbed and so result in a slower and more moderate rise in blood glucose and blood insulin." Crapo's experiments revealed that "a bowl of ice cream does almost nothing to blood glucose. Nor does a sweet potato. But a white potato or slice of whole wheat or white bread sends blood glucose soaring." Thus, nutritional advice given by physicians to diabetics for the past one hundred years is patently wrong because everyone had relied on intuition rather than experimental testing.

The second example comes from a book called *Lipids in Human Nutrition* by Germin I. Brisson, Professor of Nutrition at Université Laval in Quebec. Dr. Brisson, after a thorough review of the current literature, concludes:

> *The intake of dietary cholesterol has no significant effect on the concentration of cholesterol in the blood of healthy persons representative of the general population.* This point of view has been shared by Health and Welfare Canada since 1977 and was unequivocally adopted by the United States Food and Nutrition Board, National Research Council. Considering the slight effect that a diet high in linoleic acid and other polyunsaturated fatty acids has on serum cholesterol level, and considering the uncertainties relating to the dangers of an increased consumption of PUFA *it would seem that, for the moment, everyone concerned should be warned against all propaganda encouraging the general public to increase its consumption of linoleic acid and other polyunsaturated fatty acids.*

Twenty or more years of advice given by physicians, the American Heart Association, and innumerable television commercials are now placed in doubt. We are entitled to raise the questions as to the basis of the original advice. Were the recommendations scientific, and what was then known?

In order to shed some light on the origins of nutritional information I went on a retreat to the Sterling Medical Library, where I sat many hours poring over *Nutritional Abstracts, Series A, Human and Experimental.* This work abstracts more than nine thousand articles and books per year, thoroughly covering the field.

The entries in each issue are divided into categories. About one-fourth are in analytical biochemistry and deal with the composition of foods and techniques of analysis; even more of the abstracts deal with physiology and biochemistry. A typical study reported is on fifteen human subjects or sixty Wistar rats on controlled diets for periods ranging from days to months. While obvious effects will show up, no subtle features of nutrition can emerge from such time- and population-limited studies. The reports on health and human nutrition are mostly statistical and epidemiological investigations in relatively small populations. Data are obtained by interviews, with all the attendant limitations of that technique. The final collection of abstracts on disease and therapeutic nutrition contains reports of studies that again are based on small groups of patients or experimental animals.

What clearly emerges is the absence of long-term studies on large numbers of humans eating controlled diets and living under controlled conditions. Such studies are, in fact, not possible because of the enormous expense and the resistance of humans to being controlled experimental animals. The obvious results of metabolic biochemistry and minimum nutritional requirements can and have been established by small-scale studies. They are well known and have been incorporated into government nutritional standards. A number of clearly toxic factors, such as high sodium levels for hypertensives, also have been established.

Beyond these defined norms little is known about human nutrition that is applicable to large-scale populations. Not only is knowledge lacking, but so are the resources and methodological techniques to develop such information. In short, we are epistemologically limited in our ability to search for nutritional information.

In the absence of scientific knowledge or valid pathways to such knowledge, individuals in search of nutritional certainty must turn to faith. Where these insights come from I cannot say, but they must be regarded as religious dietary codes. My main objection is to the presentation of such doctrines as if they were science.

Perhaps some day we will develop the epistemological tools to properly understand human nutrition. I'm sure that innovation can improve the scientific approaches to these most important problems. In the meantime, I recommend that nutritional dogmas be approached with a caveat emptor and a big bowl of hot chicken soup.

Some Views
of Medicine

A Cold Couch and
a Warm Heart

For about fifteen years now I have been pleading with my muse for help in writing about a simple yet extraordinary man—Dr. Robert Salinger, pediatrician. Each time I took pencil in hand there was difficulty in finding the words to capture the essence of what I wanted to say. Then last night, as my wife, number-three son, and I were having dinner with a newfound friend who was experiencing the joy of the birth of his first grandchild, our friend began to talk of the pediatrician his family used when the children were young. Immediately we were members of the same fraternity: former patients of the good Dr. Salinger.

Our dinner companion was more eloquent than I and began to speak of the kind of greatness of a man like Salinger, who touched so many lives yet will hardly make a footnote in the history of our time. Indeed, I'm hardpressed to know where to go to find simple biographical facts I need for this write-up. Thoughts about him are best expressed by a quaint story comparing modern academia with classical China. Currently, the statement, "Mr. X is a great philosopher," will evoke the response, "What has he written?" In the classical period of China, the retort would have been "What kind of life does he lead?" In the latter sense, the man we are discussing was a true philosopher. Indeed, as our children grew up we often affectionately referred to the family pediatrician as "the old philosopher."

He first entered our lives when he came to our apartment one January morning to examine our firstborn, newly arrived home. I cannot recall who recommended this practitioner to us, but I do wish to thank that individual. The doctor walked in and said, "I hope you're not keeping this place so warm because of the baby." He was not only a no-nonsense individual, he was also not a man to waste a calorie. His office was never overheated, as was often noted when the children undressed for an examination and sat on his couch. In fact, when a group of his young patients reached high school age and discussed their doctor, they characterized him as the man with a cold couch and a warm heart.

Returning to the first meeting: He examined our new daughter, patiently discussed all questions, dispensed some words of wisdom, and left us feeling quite relaxed about caring for an infant.

There was an aura of quiet competence about this man. At the time of original acquaintance he must have been in his late fifties, but there was something ageless about him. And, indeed, I do not remember his looks changing very much over the next thirty years. He was of medium height, the white side of gray-haired, somewhat angular and athletic in appearance, and conservatively dressed in a nondescript suit adorned with a small bow tie.

Although biographical details are few, to the best of my knowledge, Robert Salinger grew up in the San Francisco Bay area of California and was a survivor of the great earthquake of 1906. Before World War I, he graduated from the University of California at Berkeley, and sometime after World War I, he graduated from Johns Hopkins University School of Medicine. Somehow, he ended up in New Haven, Connecticut, where he practiced pediatrics over two score and ten years.

The prime characteristic of his style was universal accessibility. Whenever parents were concerned enough to call, they either got through immediately or received a prompt return call. He also had a fine understanding of the various psychological outlooks of parents and geared his responses to his assessment of both parent and child.

During the period in which he took care of our children, he had been in practice long enough so that one had the impression he had seen everything. He had begun in the preantibiotic period, when one helplessly watched children succumb to rheumatic fever. He had been through the terror of the summertime polio epidemics. He had become a very fine intuitive diagnostician and could tell a good deal by simply watching a child walk across a room.

In a period when antibiotics were being overused in pediatric practice, Salinger was a therapeutic minimalist. Of course, those were the days before defensive medicine, and I doubt that he ever gave much thought to the legal aspects of his work. He could, perhaps, afford to be such a minimalist because his keen diagnostic sense told him when it was really time to act.

With his conservative concepts of therapy and his no-nonsense views, he may have been seen as something of a curmudgeon by his colleagues. I recall his once saying, "There aren't many children who need their tonsils out. There are otolaryngologists who need new cars or whose wives need new coats, but there aren't many children who need their tonsils out." That statement could hardly have made him the most popular man at the county medical society. Yet for many, many years Robert Salinger served as health officer of the Town of Woodbridge, Connecticut, maintaining the respect of that community, including the large number of physicians who made their homes there.

Our hero always thought of medicine as a modest profession, and his fees reflected that feeling. Each year, when I totaled medical expenses for tax purposes, I would marvel that even with five children our pediatrician's bills would come to around sixty dollars. Yet we never felt out of contact with this man. While it was true that we were also therapeutic minimalists, it is hard to believe how little we spent for his services. When our youngest was preparing to leave for college in 1976, he went for a required inoculation, which, I believe, was the last official service the pediatrician performed for us. "Take cash," my wife said. "The charge will be so modest I don't want them to have to bother sending a bill." We can't remember the exact fee, but I'm sure it was under five dollars, and I half remember it as being two dollars.

Although the town health officer of Woodbridge was clearly a member of the establishment, he had a contempt for meaningless paperwork. Year after year he filled out health forms for camps, schools, and colleges. Occasionally, to test the system, he would write on the form, "This child has a very serious condition. Contact me immediately on his arrival to discuss appropriate measures." The notes, of course, went unanswered, confirming his belief that the completed forms went unread.

In Amity Regional High School, which our children attended, students who were patients of Dr. Salinger felt as if they belonged to a club. They all remembered his large office, his enormous collection of stuffed animals, his relaxed manner, and his cold couch. Being patients of the old philosopher somehow established a bond among these young people.

When I heard that the doctor was about to retire, I appeared one day with a small tape recorder and asked him if he could talk about recollections of his long practice. He thought for a while and couldn't think of anything out of the ordinary that had happened. That's the way it was with him: Everyday excellence in helping people was simply nothing out of the ordinary.

The old philosopher is not a writer's dream. This man ennobled the ordinary, he lived it with vigor, and his frank talk often hid his simple eloquence. Instead of stressing the drama of medicine, he downplayed it to make it easier for people to cope. He was a quiet, reassuring presence. That doesn't sound very exciting. Nevertheless, there are thousands of people out there whose lives were in many ways made better by contact with this man. Such a nonjournalistic kind of greatness provides a resounding answer to the query, "What kind of life did he lead?"

Vital Records of Athol

I have for several years taught a course on biology and literature. It is for nonscience majors and uses literary themes to introduce various aspects of science. Thus, virology, for example, enters through a reading of those sections of Sinclair Lewis's *Arrowsmith* in which young Dr. Martin Arrowsmith discovers the mysterious bacteriophage. The last time I taught the course, the term paper assignment was to write a short story that had as a central theme some aspect of modern biology, some feature of contemporary understanding that had been gained after World War II.

Upon reading the papers, I was surprised and dismayed. The stories were mostly about cataclysms, catastrophes, and apocalypses. This was the message about science that had somehow impressed itself on the minds of these bright young liberal arts students. As one of the world's unrepentant optimists, I needed an affirmation to counter the malaise these young folks were clearly harboring.

I decided that a view of infant and child mortality as seen in a postcolonial American cemetery might convince these students that science had indeed improved our lives. "The good old days" had frightening aspects that have been all but forgotten. I was going to copy my primary data off local tombstones, but it was a chilly April day, and my friend, history buff Brooks Shepard, reminded me that the New Haven Historical Society library doubtless had the information I wanted.

Wandering through the stacks at that library, I happened on *Vital Records of Athol Massachusetts to the End of the Year 1849*. Pages 183 to 230 of that work record every

186   ATHOL DEATHS.

BISHOP, Jane, w. Tony, Nov. 27, 1804, a. 50. G.S.I.
BLAKE, Charles W., s. Thomas H. and Eunice W., Feb. 6, 1843.
    a. 5 m. G.S.2.
Ella G., d. Bradley B. and Harriet, Feb. 25, 1849, a. 8 m. G.S.I.
Fred B., s. Thomas H. and Eunice W., Oct. 1, 1845, a. 20 m. G.S.2.
BLANCHARD, Mariam, Jan. 22, 1819, a. 17 y. In a consumption,
    C.R.
——, inf. of Moses and ——, March 20, 1795. C.R.
BILLINGS, Mary Ann, d. Israel, July 21, 1842, a. 5.
Mary Ann T., d. Erastus and Abigail R. E., July 2, 1842, a. 4 y.
    6 m. G.S.2.
BLISS, Persis, w. Stephen W., Jan. 15, 1844, a. 31 y. 5 m. 1 d.
    Consumption.
Joseph W., s. Stephen W. and Persis, Sept. 5, 1838, a. 1 m. 2 d.

death in Athol, from its settlement around 1750 to 1849. Using gravestones, church records, and town records, some dedicated archivists in the early 1900s had constructed the complete necrology of Athol. The town is not unique: Records exist from many other communities. But among the vital statistics in the New Haven collection, the story of life and death in Athol seemed to contain the information I was seeking.

The first page of "Athol Deaths" begins to make the case. Of the eighteen deaths listed, six were of persons two years old or younger, two were five years old, two died in their twenties, three in their thirties, one at forty-two, one at sixty-two, and three adults at unknown ages. Three of the infant deaths were listed as caused by dysentery or teething and dysentery. The second page continues the story. Of the twenty-two persons whose ages were reported, seven died before reaching the age of three, one expired at seven, another at ten, three in their twenties, four in middle age, two in their seventies, and four in their eighties. One was listed simply as "old." The message in these pages is clear: The Grim Reaper struck early and often. Part of the original record is reproduced to help convey the perilous character of life in those times.

It is, of course, one thing to read vital statistics as abstractions and another to think about flesh-and-blood individuals. Look at the page and think about Thomas and Eunice Blake. A few months after son Charles died at age five months in February 1843, they conceived another child. And by October 1, 1845, he too was carried away by some childhood ailment. Or there is Stephen Bliss, who saw an infant son perish in 1838 and six years later watched his thirty-one-year-old wife die of consumption.

These are not unique events. Similar things happen today, but they are the individual tragedies of our time. In pre–Civil War Athol, these tragedies were the rule. When the average life span is thirty-five years, the quality of life is very different from what it is when people live seventy years. No one entirely escapes pain and suffering in any age, but the range of expectations is totally different today from that of earlier ages.

And the question I would address to the college students who see modern technology as a curse rather than a blessing is, Would you make the trade-off? Would you trade your life now for a life with the joys and pains of postcolonial America? Forget for the moment your Walkman and your word processor—would you trade immunizations, antibiotics, and a reliable water supply for freedom from air and water pollution and the threat of Armageddon?

Why, I ask, is it so much easier for my students to focus on the threat rather than the promise? Why is it so much more fashionable to lament rather than to set to work on solving the problems of our times? If I knew the answers to those questions, I would be busy teaching them. I think I need some help on this from my students.

In the meantime, I must share a brief epilogue. These words were written on a flight from Chicago to Albuquerque. The writing was interrupted by dinner, and I got into a conversation with the couple sitting on my right. In the course of the conversation they mentioned that they were from Athol, Massachusetts. It's hard to believe. I never knew anyone from Athol—indeed, never even knew that Athol existed until a few weeks earlier.

After dinner I retrieved the copy of the Athol records from among the papers in my briefcase and showed it to my neighbors. They found it hard to believe. I think they regarded me as a Magus or something worse. If these records were from Salem, Massachusetts, I probably would have been in real trouble. Anyhow, I'm grateful that as a rationalist I don't have to worry about the psychic overtones of this remarkable coincidence. As I say, as a rationalist I don't have to worry. . . .

Nihil Nimis

The phrase "substance abuse" evokes an image of deals in dark alleys, money changing hands on street corners, and shivering addicts locked in cells in the agony of withdrawal. Therefore, an article with "pyridoxine abuse" in the title comes as a surprise and evokes mixed images. One does not think of that over-the-counter chemical changing hands behind closed doors, nor does one think of a nervous, sleazy character approaching one on the street with the whispered query, "Want some vitamin B_6?"

Pyridoxine, or B_6, is an essential water-soluble vitamin required by normal adults at levels of about 2 to 4 mg a day. Until very recently, it was thought that such micronutrients were safe in any amount. But then Herbert Schaumburg and his colleagues informed us about "Sensory Neuropathy from Pyridoxine Abuse: A New Megavitamin Syndrome" (*New England Journal of Medicine* 309:445, 1983). Their review of clinical information deals with seven adults between twenty and forty-three years of age who had been taking between 2 and 6 gm a day of pyridoxine. The users all manifested loss of muscle coordination and dysfunction of the sensory nervous system, and four were severely disabled. All the patients showed improvement after withdrawal. Clinical studies clearly indicated that the vitamin at those doses is a neurotoxin. Related experiments on dogs showed that pyridoxine ingested at 0.3 gm per kilogram of body weight a day led to swaying gait, inability to walk, and selective degeneration of neurons after a relatively short time. Once again, it was shown that too much of a good thing is bad. Tablets containing 50 to 500 mg of B_6 are widely available over the counter and under a physician's advice or as a self-imposed dietary supplement; pyridoxine is probably widely abused, the more severe cases ending up in neurologists' offices.

Impelled by the slogan "It's good for you," many people
are managing to ingest toxic quantities of various naturally
occurring substances by eating very unbalanced diets or by
ingesting commercially available supplements. Nutritional
fads can, of course, be dangerous as well as foolish.

A few examples of potential food abuse help to demonstrate the problem. Many fruits and vegetables, such as
apples and eggplants, contain nonprotein cholinesterase inhibitors, which are potential dietary neurotoxins. The most
potent such inhibitors in normal foods are found in the
potato family (Solanaceae). Several cases of poisoning and
even death have been reported as a result of eating too
many potatoes, and the alkaloid solanine is suspected. The
clinical symptoms in humans and farm animals with potato
poisoning suggest neurologic as well as gastrointestinal disturbance. Populations in impoverished areas, where potatoes are the principal component in the diet, may also
display chronic neurologic disorders. Knowing all this
would not discourage me from eating potatoes in moderation, but I certainly would be wary of a fad diet based
heavily, for example, on potato skins, the part of the tuber
where solanine is concentrated. I say this having just finished a moderate and delicious lunch that included baked
potato skins and sour cream.

Another example of a hazardous food is the chickling
pea or vetch, *Lathyrus sativus*. It is widely used as a food
crop for humans and animals in many parts of the world.
It is an especially hardy plant, particularly during periods
of flood and drought, when it may become a survival food
for both humans and animals. Under such extreme conditions, the diet may be very largely *L. sativus*, and lathyrism
appears in cattle, horses, and humans. Lathyrism is characterized by spasm and paralysis of the lower legs and may
lead to paraplegia, due to neuron degeneration. The condition is irreversible. Among humans, young male adults
seem most susceptible.

A clinically well documented outbreak of lathyrism took

place during World War II among a group of Romanian Jewish men imprisoned in a forced labor camp in the Ukraine. Their daily diet consisted of 200 gm of bread and 400 gm of *L. sativus* peas cooked in salt water. After a month, painful muscle spasm and loss of bladder control resulted. After three months, paralysis of the lower extremities occurred, and the diet was changed. Some of the symptoms persisted, and a number of survivors from this camp are still disabled by spasms of the lower extremities.

The conclusion seems clear—no food or nutrient is so free of potential toxicity that substance abuse is not possible. Food faddists, ill-informed physicians, and self-proclaimed nutritionists are hazardous to your health if their recommendations contain mega-amounts of anything or fail to balance diets appropriately with a variety of foods.

The key word clearly is variety. All plants appear to have evolved alkaloids or other toxins, presumably as defenses against insects or other organisms. Since the neuromuscular junction is sensitive to a small number of transmitter molecules, it is a likely site of action for poisons. The strange similarities in structure, function, and strategy that persist throughout the animal kingdom result in some junction toxins having a wide range of susceptible organisms. If any of those animal species ingests large enough quantities of a food bearing the poison, then disability or death results.

In any case, prevention of such kinds of problems seems relatively straightforward, except in times of stress or impending starvation. A significant question remains: Why do economically well-off individuals regularly renounce moderation and engage in one or another nutritional program that leads to their own morbidity or demise? The more extreme forms of this behavior, such as anorexia nervosa and bulimia, obviously have deep psychological roots. The food fads seem somewhat related and must also be counted in the psychopathology of everyday life. On no sound scientific information whatsoever, individuals adopt unbalanced diets or supplements of such extremes that illness (and sometimes death) results. On occasion the regimens are prescribed by health care professionals; at other times they are recommended by self-appointed experts. Those angels of illness, in their zeal and hubris, pretend to knowledge that does not exist. Few things in life are more dangerous than ignorance masquerading as knowledge.

What has been lost in this nutritional madness is simple moderation in all things—embodied in the classic Latin phrase *nihil nimis,* "nothing extreme." The concept goes far back to a Greek equivalent and even further back in the biological wisdom of all those species that travel from place to place to vary their diet, acquire necessary nutrients, and avoid an excess of toxins. Much of that animal wisdom has somehow been lost to the food faddists. Too bad!

Upping the Ante

As I lay in the emergency room of Yale–New Haven Hospital with my right index finger throbbing from having its tip amputated by a slamming door, a slightly whimsical thought entered my mind. Over the years as director of undergraduate studies, chairman of the Premedical Advisory Committee, and master of Pierson College, I have written countless letters of evaluation for students applying to medical school. One of my criteria of assessment has been to address the question: If I were being wheeled into an emergency room and saw this person standing over me, how would I react? Well, my criterion was being put to a test. The cheery resident in charge came over and began to examine me. I had not known him in his premedical days. He carried out a brief physical, asked a number of questions, and as he picked up my record to start writing said, "Morowitz . . . Don't you write for . . . for . . . ?" "*Hospital Practice*," I replied, finishing his question for him. It came as a pleasant surprise to realize that someone was reading the column.

Well, here I was meeting my reading public under less than ideal circumstances, but nevertheless a certain amount of aplomb and stage presence was required. Digging deep within id and superego, I summoned up my most authoritative voice and said, "Well, what do you think?"

He thought that we should call upon Dr.——, a hand surgeon who happened to be in the hospital at that time. Not having a preferred practitioner in that specialty and knowing that one of my very own readers would certainly not give me bad advice, I concurred and decided to make careful note of my experiences as a patient so that I might gain some useful insights to pass on to others. After all, no event, regardless of how unpleasant, should be without its educational value.

Some five hours later, after I had had a chance to listen to the Yale–Holy Cross football game on radio, I was wheeled into the operating room. The surgery itself, carried out under local anesthesia, was quite an interesting experience. A lateral graft was being made from my middle finger onto the tip of my index finger. The OR staff was chatty, and as time passed, the anesthesiologist, surgeon, and I began to debate who in town made the best white-clam pizza. I suddenly withdrew from the discussion with the thought that as a matter of prudence, I should make it a policy never to disagree with the surgeon while being operated on. The next stop, the recovery room, was quite uneventful, an idle hour spent listening to the beep, beep, beep of the monitoring equipment, playing away like the refrain from a Cole Porter song. The procedure went very well, and I was released the following noon with a large cast on my hand to immobilize the index and middle fingers.

Aside from some very uncertain information about where to get white-clam pizza, I had not gained any special insights into modern health care. Actually, I was bit disappointed. There had been the initial assumption that a day in the hospital would give me a startlingly new view of the health care delivery process, but everything had been so routine that I couldn't extract a single novel thought. The most memorable part of the day was listening to my roommate and his wife in endless conversation about whether or not, in spite of a troublesome hand infection, the man should sign himself out without the doctor's approval so they wouldn't lose the money they had put up for a Las Vegas jaunt.

The insight I was seeking did eventually materialize, but it took several weeks. The bills, large and small, began to arrive, and I kept careful track of the charges as I made arrangements for payment. The fees for my initial day in the hospital, plus a subsequent two hours at the ambulatory care unit to complete the plastic surgery procedures, came to $4,731.43! That figure set off a train of thought that had lain dormant for some time.

Many years and one inflation cycle ago, I had rebelled at the frequently made statement that the human body was worth only 97¢, its elemental price; I had subsequently shown that the biochemical constituents would cost at least $6,000,000. The article referred to but left unspecified the vastly greater cost of assembling the collection of molecules into a living organism.

A new way of thinking about the problem emerged from considering the fingertip approach. According to my best guess, I had lost about 1.3 gm of tissue from the end of my digit. To restore the biologic structure of my hand to some semblance of its original state had cost $4,731.43, or $3,639.56 per gram. An 85-kg person like me therefore must minimally be evaluated at 85,000 times $3,639.56, or $309,362,600, a very satisfactory upgrade from the former lowly $6,000,000. We are not only worth our weight in gold; in replacement costs, we're worth about 300 times our weight in that precious metal.

Once again, adversity had been turned into joy. Pain and suffering had served in a poignant way to teach me our intrinsic worth. Of course, we are all priceless, but it's nevertheless instructive and uplifting on occasion to calculate the lower limit of that pricelessness in terms of the coin of the realm, in terms of some easily understood numbers.

As a postscript, note that my roommate did go on his trip to Las Vegas. The time has now come when I'm fully prepared to argue with my surgeon about pizzerias, but first I'd like to thank him for competent and thoughtful services. And finally, thinking about medical economics, my joy in rediscovering our worth was made even more complete when I found out that the $4,731.43 was entirely covered by third-party payments.

Hardware, Software

The occasion that set off this stream of thoughts was the viewing of a short movie sponsored by the university committee on the disabled. A number of events in recent years have gotten me to thinking about various problems associated with disabilities and the societal response they evoke, and this particular film dealing with the use of computers in normalizing the lives of the physically disabled struck a sympathetic chord. In computer terminology, there lies a way of looking at the world that makes disabilities somewhat easier to conceptualize and respond to in a positive manner.

The discipline of computer science is divided, conveniently but not absolutely, into the domains of hardware and software. The hardware is the chips, wires, switches, and all the other materials that go into the manufacture of computers as well as display screens and printers. Software includes the programs that direct the activities of the hardware. Programs are written on disks, tapes, and similar devices, and information is stored on hardware.

But software is different in character from hardware and nearer to what we usually associate with the activity of the mind. The software/hardware dichotomy is a contemporary example of the traditional, ever-recurring mind/body problem in philosophy. Questions arising in the computer age keep forcing us to face that difficult area of inquiry. The very phrase *artificial intelligence* brings into sharp focus the relationship between computers and minds.

Within this computer-based way of looking at the world, disabilities may generally be classified as hardware problems: in the wiring of the neural networks, in the links of the networks to the sensory or motor apparatus, or actual malfunction, loss, or developmental abnormalities somewhere in the neuromuscular apparatus. The disabilities may be congenital or the result of birth or postbirth trauma—"postbirth" extending over a lifetime.

The movie mentioned above focused on the technology that allows persons with physical disabilities to interact with a common universe of information through computers. The input to computers is generally through a keyboard, and the output is through a screen or printer. The keyboard, or its equivalent, can today be operated by fingers, toes, a pointer held in the mouth, or eye motion. Thus, almost no one has so severe a physical disability as to be denied access to computer inputs, assuming that funding is available for specialized hardware. Likewise, outputs can be visual, auditory, or tactual, so that almost no one need be denied access to computer outputs.

Since many jobs nowadays are informational and primarily require interaction with computers, a wide range of employment is open to those with disabilities. That is good news indeed. Computers will allow us to redefine "disabled" and to include in the "economically able" a large group of people who had previously been excluded.

As I was walking home from the movie on disabilities, thoughts came to mind of a recent case involving a college student with dyslexia who applied for an extra term to complete his bachelor's degree. The university has a strict rule that this degree must be completed in eight terms of residence, so the case was precedent-setting. The student, who had high standards of accomplishment, argued that background reading and organization of a substantial senior thesis would take him considerably more time than expended by the average student and that the extra term was required to complete the high-quality thesis he would like to submit.

In supporting the student's application for an exemption, I argued that there is a wide group of learning disabilities ranging from mild dyslexia to severe speech and reading impairments that probably represent congenital "hardware" abnormalities in the central nervous system. Many such persons develop their own innovative mental "software" programs to overcome their hardware deficits, and some of those people achieve outstanding success in their fields of interest. It is the responsibility of educational institutions to provide the widest range of support for those working to overcome disabilities. The university agreed and did grant the requested extension.

The hardware/software analogy seemed to make it easier for people without previous experience with learning disabilities to understand the situation, which suggests that it is a characterization worth retaining. The analogy also suggests emphasis on a computer approach to the education of the learning disabled, which is being attempted at some schools in a limited way. One idea that emerges is that if individuals overcome learning disabilities by developing their own mental software, it should be possible to design interactive computer software programs to facilitate the learning activities of disabled children. Such programs could assist students in developing their own software routes around hardware roadblocks. It is possible that a large group of individuals now classified as learning disabled could also join the "abled." In the compassionate applications of computer technology, there is no higher goal than this type of transformation.

The problem of designing the appropriate software can be approached either empirically, by working with children and computers in a learning situation, or theoretically, by studying cognitive learning, epistemology, neurobiology, and linguistics. Eventually, one hopes, the experimental and theoretical approaches will converge on some broader understanding of the learning process.

The use of computers in cognitive tasks has been designated as artificial intelligence. The connotations of that phrase have caused some serious difficulties, but its usage seems to have become fixed. One of the demands of artificial intelligence is that when we use computers to deal with abstractions such as learning and language, we are forced to concretize our ideas in order to put them into computerized modes. And as Seymour Papert has noted in his book, *Mindstorms,* ''It is this concretizing quality that has made ideas from AI (artificial intelligence) so attractive to many contemporary psychologists.'' Papert has been a pioneer in combining the ideas of Jean Piaget on learning and epistemology with computer methodology to develop the learning program LOGO. Determining how to modify such programs for the learning disabled will depend on our understanding in program language the nature of the individual disabilities. Alternatively, experimental findings on how to so modify the programs should be of conceptual value in understanding normal and hardware-altered mental activities.

I have the gut feeling that we are standing on the threshold of an era when computer technology can substantially meliorate the lives of a large segment of our society affected by disabilities of one kind or another. I'm elated by the concept of using this technology to convert the disabled into the able. The process of that conversion also provides the chance to learn a great deal about how we function at the cerebral level. Given those potential payoffs, it is difficult to resist the urge to rush out and begin work in this branch of science. Lacking the appropriate background, the best one can do is to offer support to individuals engaged in these areas of research.

Myasthenia Gravis and Arrows of Fortune

When Aristotle Socrates Onassis appeared in public late in 1974, his dark glasses hid strips of adhesive tape holding up his drooping eyelids. Those modest prosthetic devices were employed to counter the effects of the shipping tycoon's myasthenia gravis, a disease characterized by loss of muscular control in the eyelids and elsewhere. By 1974 the neurologic syndrome that afflicted one of the world's wealthiest men was rather well understood and usually quite manageable. His death a year after diagnosis of myasthenia was due to other causes, complicated by the disease of his nervous system.

The understanding and ultimate clinical control of myasthenia gravis is one of those great medical adventures that cut across modern intellectual history, involve vast areas of biology, and have vital implications for major philosophical inquiry. The tale we are about to unfold has something to say about the ways basic science and medicine interact to the ultimate benefit of all of us. It begins during the prehistory of the Indians of South and Central America. In the Orinoco and Amazon valleys and in the Guianas lived early, if unheralded, founders of toxicology. Among the most common types of biologically active preparations devised by those unrecorded natural-product chemists were extremely potent poisons used on the tips of arrows and blowgun missiles. The principal application of arrow poisons was in the hunting of animals and birds, which could be brought down by smaller, lighter, more accurate projectiles when death was by chemical rather than physical trauma. The toxic substances needed very special properties: small quantities had to produce profound and rapid effects, they had to retain their activity after drying, and they had to be readily available. To meet those criteria, the Native Americans devised extracts from a number of liana vines, usually in various mixtures with insect and snake venoms.

The most potent group of their poisons came to be called by the nonspecific name curare (*urari, wourari, curara, wourali*). Among their active ingredients are a number of highly toxic chemicals called alkaloids, a group of nitrogen-containing substances that is widely distributed throughout the plant kingdom. The historically best known is probably the hemlock that Socrates drank in fulfillment of his sentence for impiety. Alkaloids of more recent importance include morphine and nicotine. Hundreds of other biologically active members of the group are known.

Poison-tipped weapons were by no means unique to the Americas: various cultures around the world have employed a whole series of plant, reptile, and insect toxins to increase the efficacy of their weaponry. What distinguished curare was the extraordinary speed with which it totally immobilized its victims. For a number of reasons it and many other plant and animal toxins have come to play a special role in medicine. Chemically blocking an organ or cellular function provides insight into the nature of that process. Much of modern biochemistry has emerged from studying the effect of specific inhibitors—poisons that act at a known point in the chemical pathway. It has thus been possible to resolve long, complicated chemical processes into smaller, more easily understood components.

Curare was used not only in hunting but also in intertribal fighting, and it thus served as a sort of early weapon of chemical-biological warfare. The Spanish and Portuguese explorers and settlers came to fear and respect the armament of the jungle dwellers. They were especially fascinated by arrow poisons, and they periodically took or sent back to Europe samples of crude curare preparations. Such interest continued among generations of European visitors to South America. Returning from his survey of the Amazon rain forest in 1751, geographer Charles Marie de la Condamine had in his possession substantial samples of the deadly material. For the next hundred years, sporadic research took place on the effects of curare on laboratory animals.

Three types of curare were distinguished according to the containers the Indians packed them in: tube (or bamboo), pot, and gourd. The purified form of the most active component became designated by the chemical name α-tubocurarine chloride, as a reminder of its original packaging.

By the late 1840s curare came to the attention of Claude Bernard, who had come to Paris in 1834 hoping to be a playwright. The literary critic Girardin read his script and advised him to go to medical school. During the next ten years, while supporting his research with the dowry from an arranged and unhappy marriage, Bernard laid the foundations of experimental medicine. Curiously enough, while the physiologist was working out the theory and technology of modern experimentation with animals, his disaffected wife and daughter were active in the French antivivisection movement.

Experimenting on frogs, Bernard and his colleague M. Pelouze verified that curare in the bloodstream causes paralysis throughout an animal's entire body. Then they devised a simple yet elegant series of experiments on frogs that demonstrated that curare acts by blocking the transmission of signals from the central nervous system to the muscles. By restricting the flow of blood (and hence curare) to a frog's hind leg and then determining the neural response, the French scientists identified the neuromuscular junction as the site of action of curare. They thus were able to conclude that the victims of a curare-poisoned arrow experience the terror produced by the combination of a normally working sensory nervous system, a totally unimpaired mind, and the absolute inability to move a single muscle. When enough toxin reaches the respiratory neuromuscular junctions, breathing stops and death ensues. If artificial respiration is used until the curare is metabolized and eliminated, survival is possible.

Because curare was able to relax muscles in contractive spasms, such as occur in lockjaw, scientists looked for some safe way to administer the alkaloid locally or in sublethal doses. In any case, Bernard's results made it possible to distinguish neuromuscular disorders from primary muscular disorders and were thus of major diagnostic value. While studies on curare were proceeding among physiologists, clinicians were becoming familiar with a somewhat diffuse disease characterized by weakness and fatigability of the skeletal muscles. The clinical features had been noted in 1671 by Thomas Willis, one of the foremost British physicians of his time; he is also credited with rediscovering the diagnosis of diabetes mellitus by the sweet taste of diabetic urine, a technique that had been lost since the time of Galen. By 1900 the myasthenia gravis syndrome had been thoroughly described and was well known to the medical profession. During the early part of this century it was sometimes called Goldflam's disease or Hoppe-Goldflam disease, after Swiss physiologist Johann Hoppe and Polish neurologist Samuel Goldflam.

In 1901 the physiologic effects of curare and the nature of myasthenia gravis came together when neurologist Hermann Oppenheim pointed out the similarity between the symptoms of curare poisoning and those of myasthenia gravis. If curare acted at the neuromuscular junction, might not myasthenia gravis be a disorder at the same location? Because curare was such a universal poison, affecting all the animals on which it was tried, scientists had no difficulty in moving from animal studies to ideas about human disease. In his 1865 book, *Introduction à l'étude de la médecine expérimentale*, Bernard had provided the rationale for such mental leaps.

The contribution of curare to understanding myasthenia gravis, important though it was, constituted just one element in solving a complex interlocking puzzle. Another important component was understanding the mechanism by which impulses are carried by nerves. The modern study of the electrical excitability of nerves, the transfer of signals to muscles, and the beginnings of electrochemistry itself go back to the experiments of Italian physician Luigi Galvani in the late 1700s. Much early electrical research emerged from electrophysiology, and in the early days a nerve-muscle preparation (the twitching frog's leg) was the most sensitive available device for electrical measurement. One of the great scientific disputes of all time, on the interpretation of animal electricity, ensued between Galvani and physicist Alessandro Volta. (We honor Galvani with, among others, the word "galvanometer," and Volta, with "volt.")

Without detailing that dispute we note that as instruments improved, it eventually became clear that the excitatory process in nerves consisted of a propagated disturbance, which was most easily measured by changes in voltage across the neural membrane. It took another two hundred years for us to fully understand the nature of the electrical events accompanying the transmission of a signal along the axon. The biophysics of the action potential was finally elucidated in the North American squid. Traversing almost the entire squid body, which may be over twelve inches in length, are giant axons, so called not because of their length (long neural axons are common) but because of their large diameter, which permits insertion of internal electrodes. As often happens, experimenters had found a biologic system ideally suited for the experiments they wished to carry out. And so every May, June, and July—the squid season—neurobiologists from around the world gathered at the Marine Biological Laboratory in Woods Hole on Cape Cod to carry out their experiments. The giant axon indeed proved to be an ideal system for precise electrical measurements, which led to a series of powerful theoretical constructs and to the understanding of axon physiology and signal transmission in all species with neurons.

Regarding events at the neuromuscular junction, it has been known since 1900 that both curare and myasthenia gravis involve a blockage at this site and that the signal carried through the nervous system to the neuromuscular junction is electrical in nature. Two schools of thought existed. Members of the "soup" school believed that chemicals were released on the nerve side of the synapse, taken up on the muscle side, and used to activate electrical events in the muscle. "Spark" school proponents believed in a direct transfer of electrical excitation from nerve to muscle. The next big breakthrough was twenty-five years in coming.

The mystery of the junction began to resolve itself during a series of experiments carried out in the laboratory of physiologist Otto Loewi in Germany from 1920 through 1926. Loewi worked with an excised frog heart, which he was able to keep functioning in a salt solution for a rather long time. He could either excite or inhibit the heartbeat by stimulating different sets of nerves going into that organ. If the heart was removed after stimulation and another placed in the same solution, the second heart's response would be identical to that of the first. The original nerve excitation had released into the solution some chemical substance that affected the unstimulated second heart.

Loewi and his co-worker E. Navratil demonstrated that the substance released on stimulation of the vagus nerve was acetylcholine. At about the same time, a number of other researchers discovered in the blood and muscle of many animals an enzyme (acetylcholinesterase) that rapidly catalyzes the breakdown of the released substance into acetic acid and choline. Still others showed that a frequently used physiologically active substance, eserine, operates by inhibiting the action of acetylcholinesterase.

Eserine (or physostigmine) is another of those valuable

medical substances whose source goes back to a jungle vine. The legume *Physostigma venenosum* grows in Africa and produces a toxic seed known as the Calabar bean, named after the town or river in Nigeria. At the same time that Claude Bernard was studying curare, another French physiologist, A. Vee, was extracting from the Calabar bean an active substance with powerful physiologic effects, which he called eserine. By the end of the century it had entered the pharmacopoeia as an antitetanic and miotic. The 1899 *Merck's Manual of the Materia Medica* recommends eserine for the treatment of paralysis but doesn't specify whether the paralysis of myasthenia gravis is intended.

Although some investigators suspected that acetylcholine was the neurotransmitter substance, no experimental proof was available for many years. That discovery might have been made in Germany. One day in 1933, shortly after Hitler came to power, the director of the Berlin Research Institute where Wilhelm Feldberg worked called the young researcher into his office. Feldberg had been dismissed; he was required to leave the building by midnight of that day, and he was forbidden to enter the premises again. Fortunately for Feldberg and for neurophysiology, a representative of the Rockefeller Foundation, who was in Germany to help dismissed scientists, appeared on the scene with an invitation for Feldberg to work with the distinguished Sir Henry Dale at his laboratory in Hampstead, England.

The next three years were crucial in the development of neurophysiology, for Feldberg took with him a method of quantitatively measuring acetylcholine via its biologic effect on a muscle preparation dissected out of the lowly leech. Dale, Feldberg, and co-workers established that the electrical signal in the motor neuron causes the release of a pulse of acetylcholine on the nerve side of the nerve-muscle synapse. The acetylcholine is taken up by specific receptor molecules on the muscle side of the junction, and the muscle contracts. Next, acetylcholine is broken down by its degradative enzyme and reassembled in the membranes of the nerve cells, which are now ready to fire again. Most of those experiments were carried out on dogs and cats; Dale and his colleague H. W. Dudley had done their earlier work, showing the normal occurrence of acetylcholine in the animal body, on the spleens of horses and oxen.

On May 12, 1934, before a meeting of the Physiological Society, Dale and Feldberg presented their first experimental evidence on the role of acetylcholine at the neuromuscular junction of the cat's tongue. On the very same day, Mary Walker, a resident physician, submitted a letter to the editor of *The Lancet* reporting "striking though temporary" relief of the symptoms of a myasthenia gravis patient given an injection of eserine. Laboratory and clinical studies thus joined again, opening the door to a further series of investigations and to ultimate understanding of the disease. The question of whether the neuromuscular transfer was, at the most detailed level, primarily chemical or electrical persisted, however, for several years. Those years witnessed great advances in the measurement of electrical events in nerves, the obtaining of single nerve-muscle fiber preparations, and the coming together of electrophysiologic and pharmacologic evidence.

The electron microscope and neurobiologic techniques revealed the detailed structures on the nerve and muscle membranes at the synapse and indicated that acetylcholine is released in bursts, small packages of molecules that open from the neural side into the intermembrane space at the synapse. In 1936 Dale and Loewi were jointly awarded the Nobel prize for their roles in elucidating the neuromuscular junction.

Fundamental understanding of the transmission process made it possible to reexamine the disease mechanisms of myasthenia gravis. Malfunction of the neuromuscular junction could be caused by a defect of acetylcholine production and release, a defect of acetylcholine receptors and

activators on the muscle side, or a defect of acetylcholinesterase, the enzyme that recycles the transmitter molecule.
The first task was to distinguish between those three
effects.

The clue unexpectedly came from originally entirely unrelated studies on the nature of the action of snake venom.
C. C. Chang and C. Y. Lee were working at the Taiwan
National University College of Medicine, where they were
studying the venom of the many-banded Taiwan krait,
Bungarus multicinctus. One of the most toxic of all snake
venoms, it is at least ten times more lethal than cobra
venom. The reptile is native to Burma, southern China, and
Taiwan, where bites are regularly reported. Chang and Lee
found that the venom has several neurotoxic components,
including α-bungarotoxin (which acts like curare) and β-
bungarotoxin (which acts like botulinus toxin from spoiled
food, blocking the release of acetylcholine). Thus, it simul-
taneously blocks both release and uptake of acetylcholine
at the neuromuscular junction, a strategy designed for the
rapid demise of whatever is bitten by this animal—a case
of overkill, indeed. Since the food of most kraits is other
snakes, that strategy may, however, have a high payoff in
nature.

Further efforts yielded a highly purified, potent α-bungarotoxin, which attaches like glue to the acetylcholine receptors at the neuromuscular junction. (Indeed, as we have
noted, the blocking of the neuromuscular junction is the
chief reason the venom is so toxic.) By measuring how
much radioactivity sticks to junction preparations that
have been treated with the labeled snake toxin, we can
determine the number of acetylcholine receptors. Investigators at Johns Hopkins University and at the Carnegie Institution of Washington demonstrated that myasthenia
gravis patients have many fewer such receptors per cell
than do normal subjects. The techniques for those binding
experiments were first worked out on chick muscle cells in
culture.

Other investigators, following a slightly different trail in order to find an animal disease resembling myasthenia gravis, used α-cobra toxin, which acts similarly to α-bungarotoxin. Since cobras and kraits belong to related genera, similarities in their toxins are not unexpected. S. Satyamurti and his co-workers injected tiny quantities of cobra poison into rats, and the experimental rodents promptly showed all the characteristic features of human myasthenia gravis. Those α-cobra toxin studies strongly supported the idea that a blocking or reduction of available acetylcholine receptor sites on muscle fibers accounts for the clinical and physiologic effects seen in human myasthenia gravis.

After establishing the mechanisms of faulty neuromuscular transmission in afflicted persons, investigators turned to the question of what causes the reduction in number of neurotransmitter receptors. Myasthenia gravis had long been suspected of being an autoimmune disease because it was associated with other defects of the immune system. Autoimmune diseases are a strange molecular-level example of the problem of distinguishing self from nonself. To effectively combat infectious diseases, the body's immune system stands poised ready to manufacture antibodies to any foreign macromolecule, virus, or bacterium. The more ready the system is to mount an effective defense, the more likely it is to make a mistake and treat a constituent structure as a stranger. An analogy can be seen in poised police and military defense systems, which may respond in the absence of a real threat. In any case, much human disease, particularly pathology associated with aging, is now believed to be caused by such recognition errors of the immune system. The reason for the failures is under intensive study, and solving those problems would be an enormous leap forward in maintaining human health and well-being to an advanced age.

In the case of myasthenia gravis, researchers postulated that antibodies to the neuromuscular acetylcholine receptors were causing the disease. In one series of passive transfer experiments, antibodies were purified from the blood of myasthenia patients and administered to mice. The recipient animals developed symptoms of myasthenia gravis, presumably caused by the presence of the acetylcholine-binding antibodies in the human blood protein preparation. By several indirect procedures, antibodies to receptors were found, but a closer animal model of the disease was needed to prove the explanation of the cause. Once again, seemingly unrelated findings provided a path to the solution of the problem of interest.

This time the study goes back over 2,300 years to Aristotle, the father of biology, who wrote on the properties of electric fish. The highest voltage and largest current produced by such animals are found in the electric eel, *Electrophorus electricus* (really not an eel but a fish, according to taxonomists), whose electric organ, which packs a wallop of 300 to 600 volts, evolved from muscle tissue (as did all the electric organs of fish as well as those of the electric ray, or torpedo). Being derived from contractile tissue, the electric organs receive their messages to fire from the central nervous system by synapses that are similar to neuromuscular junctions.

In the 1950s Carlos Chagas and his co-workers in Rio de Janeiro began the isolation and purification of acetylcholine receptors from Amazon River Electrophorus. By the 1960s they had obtained rather pure preparations. (Curiously enough, one of the stages of purification consists of reacting the soluble protein with concentrated curare, which by binding to the protein renders it insoluble.) When the investigators injected the purified acetylcholine receptor molecules into rabbits, the animals mobilized their immune systems and began to make antibodies to the foreign protein. The antibodies also attacked the receptor molecules of the rabbits' neuromuscular junctions, and the animals began to display symptoms of myasthenia gravis.

Those experiments provided an immunologic model of the disease. The procedure has a number of subtleties. The rabbit immune system has to recognize the electric-organ acetylcholine receptor as a foreign protein, yet the strange molecules must be sufficiently similar to the rabbit's own acetylcholine receptors that antibodies to the former will attach to the latter and mimic the disease. Just the right amount of difference is required, and once again nature has been kind to the experimenter, for precisely that degree of difference exists.

Thus, many lines of evidence came together to confirm that autoimmune attack on acetylcholine receptors at the neuromuscular junction is the immediate cause of myasthenia gravis. Note that in this brief sketch of how researchers managed to put together this fascinating story, we have dealt with experiments on liana vines, Calabar beans, frogs, squids, leeches, dogs, cats, horses, oxen, kraits, cobras, rats, chickens, electric eels, electric rays, rabbits, mice, and humans. A more detailed historical record of the neuromuscular junction would certainly add other species to that impressive array of plants and animals. We are surely seeing an exquisite example of biologists' skill in developing appropriate experimental systems and relating the results from taxonomically diverse species.

We next ask the question, Why is it possible to transfer information with seeming ease across such broadly divergent groups of organisms? The obvious evolutionary answer of common ancestry is not quite satisfying, for the physiologic similarities seem greater than we might expect. The neuromuscular junction seems to have undergone very little divergence through the past 600 or 700 million years or more. Lower organisms having evolved the acetylcholine-releasing, -binding, -signaling, -hydrolyzing, -regulating system, little change was apparently necessary in mammalian development. From such a starting point for nerve-muscle signaling, it may be nearly impossible by evolutionary pathways to arrive at anything better. It is not always the case that physiologic information can be so easily applied among widely separated groups of animals, but when it happens, those examples give us a wonderful sense of the underlying unity and continuity of all life. The great importance of toxicologic agents in the study of myasthenia gravis leads us to inquire why plants have evolved so many animal poisons.

It seems clear that liana vines have not evolved the synthesis of α-tubocurarine, the major chemical toxin in curare, in order to poison arrow tips, but one can see the long-term advantage of a plant's possessing animal poisons. If the point of attack is the neuromuscular junction of the plant eater, which would be a very good characteristic of a natural insecticide, then the molecules synthesized by the plants will also affect humans and other animals. (Similarly, industrial and university chemists have developed some insecticides that we have had to refrain from using because they are too toxic to humans and other animals.) The strong, evolution-based chemical similarities among living organisms make it difficult to devise highly specific poisons that are directed at just a few species.

To inhibit one set of organisms and not another requires selective strategies. A good example is penicillin, which interferes with the synthesis of cell wall in bacteria. Since the mold that produces the antibiotic and the humans who use it lack molecular structures chemically similar to the bacterial cell wall, penicillin is relatively nontoxic in its medical applications and to the species that synthesize it. With the neuromuscular junction, selectivity is very difficult, because the biochemical mechanisms, cellular structures, and strategies have remained so similar throughout evolutionary history. Thus, in retrospect it becomes clear why research on such diverse living things has, in this case, come together to provide an understanding and therapeutic rationale for a human disease.

It has been fashionable from time to time to chide scientists and federal granting agencies for research that seems remote from our practical problems. Indeed, a group of legislators has made a parlor game out of laughing at the strange titles of some research projects. The understanding of myasthenia gravis presents the other side of that story. Scientifically valid research carried out to answer basic unresolved questions has, in spite of its seemingly obscure character, provided the information needed by the clinician to deal with real patients with real illnesses. All intellectually valid biologic and biochemical research work has the potential of filling in pieces of a puzzle, segments of which will be of value in socially significant ways. Jigsaw and crossword puzzles are good analogies; every piece properly placed or every word properly solved relates to the whole and contributes to the solution of the rest of the puzzle. Our knowledge of biology is, as modeled by these puzzles, one vast, interconnected network held together by common biochemical mechanisms, cellular structures, physiologic processes, and organismic strategies. The connectedness of the network extends knowledge obtained about one species and allows it to spread out and be amplified and amended as it is applied across the relevant taxa. It provides another example of the quasi-religious idea of the interrelatedness of all things.

No research program by a hypothetical National Institute of Myasthenia Gravis one hundred years ago could have, before the fact, extended its vision to the very broad range of animal experiments and systems required to understand the disease. In biologic research the shortest distance is often not a straight line, and we usually lack the wisdom to foresee the exact path. In such a world, understanding the fundamentals of biology and biochemistry is often the best route to success. Information moves ever more quickly from the laboratory and field to the clinic and treatment facility, but we are still often dependent on the intuition and creativity of investigators.

As a result of our understanding the basic mechanisms of myasthenia gravis, rational therapy has become possible. Daniel B. Drachman of Johns Hopkins University School of Medicine writes, ''Modern treatment has improved the outlook in MG so remarkably that most patients are able to return to a full productive life. The ultimate goal of a cure, however, has remained elusive; most patients must continue taking medication despite the risks of adverse side effects."

The four current modes of treatment are improving signal transmission at the neuromuscular junction, suppressing the immune system, removing the thymus, and decreasing circulating antibodies (see R. P. Lisak, "Myasthenia Gravis: Mechanisms and Management," *Hospital Practice*, March 1983). The first therapeutic mode goes back to Walker's original use of acetylcholinesterase inhibitors, such as eserine. A number of agents are now available that delay the breakdown of the transmitter molecules, thus improving the operation of the junction. Experiments are also under way on drugs that increase the release of acetylcholine on the nerve side of the junction. Immunosuppression is now used in treating a wide variety of diseases, and a number of immunosuppressive drugs are available. Several are used in myasthenia gravis patients. The thymus may be the principal site of production of antibodies to the acetylcholine receptor, or the site of antigenic stimulation, or the site of disturbed immune regulation. In such cases, surgical removal of the thymus frequently leads to improvement in some and complete remission in others. Reduction of circulating antibody levels by withdrawal of lymph via the thoracic duct or by plasma exchange is a temporary expedient to get patients through especially bad periods.

Research is in progress to specifically suppress the production or inhibit the action of the antibodies responsible for myasthenia gravis. The basic understanding of the neuromuscular junction and the immune system gained during one hundred years of research makes possible a fully rational approach to the search for a permanent cure.

Thoughts about biologic and medical aspects of the neuromuscular junction suggest some philosophical implications. Since the time of Plato, if not earlier, philosophers have dealt with the mind/body dualism. The issue was formalized in 1630 by René Descartes, who postulated the existence of material substance and mental substance—the latter a property of human beings only. Only man "doubts, understands, conceives, affirms, denies, wills, refuses, . . . imagines and feels." The great French philosopher left unresolved the relation between matter and mental substance. Science has dealt largely with matter, and indeed, during the seventeenth through nineteenth centuries scientists systematically tried to ignore the mind. Philosophers maintained four divergent viewpoints: only mind is real, and individuals' minds are connected by the mind of God (George Berkeley); only matter is real, and mind is an epiphenomenon (the behaviorists); mind and matter both exist (Descartes); mind and matter are both aspects of the same underlying reality (Spinoza).

As the twentieth century has unfolded, scientists have found it harder and harder to ignore the mind. Relativity theory was formulated around the notion of an observer. Quantum mechanics demanded the mind of an observer to move from a hypothetical probability distribution to a measurement. Finally, in thermodynamics, the last stronghold of nineteenth-century physics, it was shown that entropy is a measure of the observer's ignorance of the exact state of the system; it is a property of the mind of the observer.

With the mind intruding on biology from the underlying science of physics, it has been necessary for biologists to once again think about the mind, which is closely associated with the brain and neural networks. Within this context there has been a philosophical focus on the neuromuscular junction. For dualists, who would separate mind and body, the neuromuscular junction is the chief interface between those two aspects of existence. A willful act is associated with action potentials in the central nervous system; then the neuromuscular junction is crossed, and a muscle action ensues. The neuromuscular junction is the point at which volition becomes action. For a person with a sublethal dose of curare, no action is possible no matter how strong the will. To the dualist, curare is a reagent that separates the body from aspects of the mind. The mind/body problem has traditionally been the sole province of philosophers, but as neurophysiology and muscle physiology present a more and more detailed picture of the cellular and molecular events accompanying thought and action, we may have to include that understanding in our philosophy.

Our story has gone from the rain forests of South America to the laboratories of Europe, the Americas, Asia, and Australia, to clinics around the world, to the philosopher's study. The neuromuscular junction connects many aspects of our humanity—or, at a deeper level, it connects many aspects of our animality. Knowledge of the junction enables us to deal with a number of diseases and may open the door to a new understanding of ourselves.

Some Usual
and
Some Unusual
People

Fathers and Sons

It was a small inside-page obituary headlined "100 Zen Students Keeping Vigil over Body of Man Slain in Streets." The San Francisco dateline drew me to the article, and the discovery that the stabbing victim was twenty-three-year-old Chris Pirsig sent me into a melancholy reverie. For those of us who had been drawn into the maelstrom of madness related in *Zen and the Art of Motorcycle Maintenance*, Chris was no literary abstraction but a flesh-and-blood individual. This was not the report of the death of a stranger but of the young boy who had been the ultimate reason for Robert Pirsig's coming out of the mental hospital, for "to have let him grow up alone would have been really wrong." So it was that four of us took that motorcycle ride from Minnesota to Bozeman, Montana: Robert and Chris Pirsig, the ghost of Phaedrus, and I, the reader, trying to see reality and meaning through another's eyes.

Robert Pirsig, the central character in this autobiographical story, was returning to Bozeman to relive, or to exorcise the memory of, events that had driven him to a mental hospital. He brought his son on the trip in an effort to deal with early symptoms of instability that he saw developing in the boy. He felt that these might have been the result of Chris's trauma in seeing his father institutionalized. Phaedrus was the author's name for himself before the fateful day when he came apart at the seams. Robert was trying to face the past through Phaedrus and the future through Chris.

Pirsig's book, among its many levels, is a father-and-son story, a subject which seems to be of little interest to contemporary writers. The shock of the Oedipal theme has passed, and the Victorian conflict between the rigid parent and the rebelling young man has become passé. Yet *Zen* had another theme; for without being maudlin it talked of love between a father and a son, and the fourth rider on the motorcycle came to sense the value of that emotion. It is all right to love one's father and one's sons. That is an important thought, particularly in these days of so many fatherless families. It leads one to question why the government fosters this institution among the poor and why a failure of commitment between the sexes abets the practice among the more affluent. No amount of rational discussion of the paternal bond can have the impact of Robert's realization that "I haven't been carrying him at all. He's been carrying me."

A rereading of *Zen and the Art of Motorcycle Maintenance* brings us to feel the full wrenching pain of Chris's death. The following conversation takes place between father and son, who are camping out together:

"Dad, what do you think about all the time? You're always thinking all the time."
"Ohhhhhh . . . all kinds of things."
"What about?"
"Oh, about the rain, and about troubles that can happen, and about things in general."
"What things?"
"Oh, about what it's going to be like for you when you grow up."
He's interested. "What's it going to be like?"
But there's a slight ego gleam in his eyes as he asks this and the answer as a result comes out masked.
"I don't know," I say, "it's just what I think about."

The theme of Chris's maturing recurs in another conversation between these two at a later time and place.

"Dad?"
"What?" A small bird rises from a tree in front of us.
"What should I be when I grow up?"

The bird disappears over a far ridge. I don't know what to say. "Honest," I finally say.

"I mean what kind of a job?"

"Any kind."

"Why do you get mad when I ask that?"

"I'm not mad . . . I just think . . . I don't know . . . I'm just too tired to think. . . . It doesn't matter what you do."

Of course, it does matter; Robert was not being honest. All parents are concerned with what their offspring will do. Children are the one unquestionable tie with immortality; all else is faith or hope.

A mark of great writing is the extent to which it takes us beyond ourselves and involves us so deeply in the lives of the characters that our emotions become entwined with theirs. Pirsig has succeeded too well, and now we must also bear the pain. Like Chris's family and friends we must try to lift ourselves from the abyss of gloom and move on.

The short newspaper article omits, in its terseness, what had intervened in the life of Chris Pirsig from the questioning young boy to the mature student of Zen meditation. Robert Pirsig's book ends with the note:

Trials never end; of course, unhappiness and misfortune are bound to occur as long as people live, but there is a feeling now, that was not here before, and is not just on the surface of things, but penetrates all the way through: We've won it. It's going to get better now. You can sort of tell these things.

The actual end of the Zen journey on the streets of San Francisco is one that few novelists would dare to have written. The apparent meaninglessness of the manner of Chris's death, capriciously knifed while walking a few blocks from home, sends us reeling and seeking for meaning. It is a koan, or Zen riddle, without an answer and without a master to lead us to understanding. *What is Quality?—Chris is knifed.* We are forced to look for comprehension within ourselves, a task that may take the most rational among us Phaedrus-like to the edge of madness.

There is no end to this reverie; for all conclusions tend to trivialize that which has transpired between a son and a father whose gifted pen has made us all part of this drama.

Thoreau as
a Scientist

We don't usually think of Henry David Thoreau as a space biologist, yet through the pages of *Walden or, Life in the Woods,* the idea of life in the universe keeps appearing. The book was published in 1854, before the break with the view that the universe existed solely for man on earth, yet its author had a cosmic view of intelligent beings that extended far beyond his pond or planet. Well aware of the theme of Newtonian astronomy that "this whole earth which we inhabit is but a point in space," he speculated on life in the Pleiades or "behind . . . Cassiopeia's Chair."

We tend to regard Thoreau as a rustic because he tested himself in the isolation of a small Massachusetts pond and spent almost his entire life within a few miles of Concord. He writes: "Where I lived was as far off as many a region viewed nightly by astronomers." This is a pretentious statement. His remoteness was never as great as he imagined, and his thoughts were closely associated with those of his age. Yet his thinking and that of his neighbors were, on occasion, light-years apart. One of Thoreau's major roles was as critic of American society, and a commentator can never distance himself too far from those he castigates.

What keeps recurring in Thoreau's writings is speculation about beings on other celestial bodies and in more distant parts of the universe. He notes that other planets are illuminated and warmed by the same sun, and he ponders "different beings in the various mansions of the universe" who are contemplating the same star.

He presents a cosmic triangle: the distant star at one vertex, himself at the second, and some intelligent exobiologic being at the third. In the chapter "Solitude," he asks, "How far apart, think you, dwell the two most distant inhabitants of yonder star, the breadth of whose disk cannot be appreciated by our instruments?" More than most men of his age, Thoreau was aware that he lived on a small planet in a very large universe inhabited by other intelligent beings.

Unlike his contemporary, Walt Whitman, who became "tired and sick" at the astronomer's lecture and "looked up in perfect silence at the stars," Thoreau was able to combine scientific knowledge with the poet's eye. It is the ability to combine all knowledge into a coherent view that allows the author of *Walden* to speak to later generations who reach out and try to communicate with distant beings.

If we don't ordinarily think of Thoreau as an exobiologist, still less do we think of him as a biophysicist. A close reading of *Walden* reveals a rather surprising knowledge of this area.

According to Liebig, man's body is a stove, and food the fuel which keeps up the internal combustion in the lungs. In cold weather we eat more, in warm less. The animal heat is the result of a slow combustion, and disease and death take place when this is too rapid; or for want of fuel, or from some defect in the draught, the fire goes out.

Justus von Liebig (1803–1873) was a German scientist who combined results from heat measurements of living animals with the emerging insights of thermodynamics to develop the concept of *animal heat*. The work discussed by Thoreau was at the cutting edge, the forefront of mid-nineteenth-century science. One wonders how the word traveled so quickly from Giessen to Concord.

Thoreau went on: "The grand necessity, then, for our bodies, is to keep warm, to keep the vital heat in us." Having announced that principle, he finds it necessary to criticize the "luxuriously rich [who] are not simply kept

comfortably warm, but unnaturally hot.'' Each fact of nature was for him a point of departure to comment on man
and society. He concluded with the enigmatic question:
''How can a man be a philosopher and not maintain his
vital heat by better methods than other men?''

The discussion of animal heat reveals the breadth of Thoreau's erudition, as he quotes another famous scientist:

> Darwin, the naturalist, says of the inhabitants of Tierra del
> Fuego, that while his own party, who were well clothed and
> sitting close to the fire, were far from too warm, these naked
> savages, who were farther off, were observed, to his great
> surprise, ''to be steaming with perspiration at undergoing
> such a roasting.''

What is surprising about this reference is that by 1854,
five years before Darwin had achieved fame for his *Origin
of Species,* his much less famous *Journal of the Voyage of
the Beagle* was read in the apparent isolation of Concord.
Thoreau's tastes reveal a range and depth of interest in
science beyond the casual. He returns to the theme of biologic energetics in ''The Bean-Field.'' In a much more poetic mood he points out:

> We are wont to forget that the sun looks on our cultivated
> fields and on the prairies and forests without distinction. They
> all reflect and absorb his rays alike, and the former make but
> a small part of the glorious picture which he beholds in his
> daily course.

For all Thoreau's erudition in theoretical areas, the image that persists most from reading *Walden* is that of a
field biologist, natural historian, ecologist, limnologist, and
ornithologist, who has seen and felt and smelled and
touched the world of nature. This was the description by
Emerson in his eulogy to Thoreau:

> It was a pleasure and a privilege to walk with him. He knew

the country like a fox or a bird, and passed through it as
freely by paths of his own. He knew every track in the snow
or on the ground and what creature had taken the path
before him.

This was the man who in the spring and summer of 1847
collected specimens of fish and turtles for Louis Agassiz,
the distinguished Harvard naturalist. Indeed there seems
to have been an ongoing communication between Thoreau
and Agassiz during the long period when *Walden* took
shape as a book. In 1850, Thoreau was elected a corresponding member of the Boston Society of Natural History.
He had presented them with a rare goshawk. The membership gave him access to the society's library, a privilege
that he exercised over the years.

Given Thoreau's association with biologists, it is not
strange to find him using the genus and species system of
referring to the plants and animals he observed. His use of
species names was very sporadic; he always used the popular name but sometimes included the binary term. He (or
his editor) was rather relaxed about capitalizing genus
names, which is demanded by taxonomists. Thus, purslane
is called *Portulaca oleracea,* while sumac is designated
rhus globra. At times Thoreau can be very exacting about
such matters. In discussing pickerel he notes that there
was

another, golden-colored, and shaped like the last, but
peppered on the sides with small dark brown or black spots,
intermixed with a few faint blood-red ones, very much like a
trout. The specific name *reticulatus* would not apply to this;
it should be *guttatus* rather.

This excursion into systematics comes as a surprise, but
at various other points Thoreau reveals his acquaintance
with more academic biology. In discussing the eating habits
of larvae and adult insects, he refers to "Kirby and
Spence," which was the 1846 American edition of *Intro-*

duction to Entomology by William Kirby and William Spence. Thoreau's association with the Harvard biology department and the Boston Society of Natural History provided him with contacts with up-to-date scientific information.

One of the aspects of Thoreau's approach was his ability to deal with technical matters at one moment and then to turn to the practical problems of, say, catching fish. He was equally at ease with Agassiz of Harvard and with the Canadian woodchopper he frequently met near his home. And because he looked so deep within, he could extend his external gaze broadly on the world.

For all his interest in science, Thoreau was a severe critic of the way it was taught; instruction was far too theoretical and removed from the art of life. He gives an example: "To my astonishment I was informed on leaving college that I had studied navigation!—Why, if I had taken one turn down the harbor, I should have known more about it."

Threading through *Walden* is a constant attack on technology. The target may be the railroad, the telegraph, or the factory system, but the judgment is always that the products of the Industrial Revolution hamper rather than help those in search of self-knowledge and enlightenment. Much of the book is an assault by the author on the values of his neighbors, and those values include the attachment to material possessions that has come to dominate American culture. Thoreau was one of the first to point out the spiritual problems of a society built on goods made available by the Industrial Revolution. There is a difference between Thoreau and many contemporary critics. His rejection of the manufacturing industry was more complete; he did not make use of the goods and services he found objectionable. The life was consistent with the philosophy.

For the present-day naturalist, *Walden* is a rich lode from which one can extract descriptions of the plants and animals, indeed of the ecosystems, that surrounded the small New England pond. For example:

As I sit at my window this summer afternoon, hawks are circling about my clearing; the tantivy of wild pigeons, flying by twos and threes athwart my view, or perching restless on the white pine boughs behind my house, gives a voice to the air; a fish hawk dimples the glassy surface of the pond and brings up a fish; a mink steals out of the marsh before my door and seizes a frog by the shore; the sedge is bending under the weight of the reed-birds flitting hither and thither; and for the last half-hour I have heard the rattle of railroad cars . . . conveying travellers from Boston to the country.

Thoreau's keen eye also looked into the soil, where he uncovered small implements of war and hunting left by "unchronicled nations who in primeval years lived under these heavens." The artifacts of Indian civilization were there as a reminder that the woods represent change and variation. And so the philosopher-farmer became archaeologist, as his hoe tickled against arrowheads and stone axes.

"The Ponds" is probably of greatest value to current science. The descriptions of the color of the water, its transparency, the flow between the ponds, the rise and fall of the water level, and the population of flora and fauna are all limnologic points of importance. Although Thoreau's instruments were limited to a measuring tape and a thermometer, his observations present a clear view of the pond and its annual cycle. The details about Walden Pond and neighboring bodies of water provide useful data about seasonal and longer-term changes in hydrologic systems.

In "Brute Neighbors" we read of a major battle between armies of red and black ants. For a fascinating few pages Thoreau describes in vivid prose "the legions of these Myrmidons . . . engaged in deadly combat." Those lines give insight into why Thoreau the universalist was not really a scientist. Rather than the detailed descriptions that characterize the writings of such natural historians as Maeterlinck, who focused on ant colonies and observed them for long periods of time, the style of *Walden* quickly transforms the report of the battle into something far more anthropocentric:

For numbers and for carnage it was an Austerlitz or Dresden, Concord fight! Two killed on the patriots' side, and Luther Blanchard wounded! Why here every ant was a Buttrick,— "Fire! for God's sake fire!"—and thousands shared the fate of Davis and Hosmer. There was not one hireling there. I have no doubt that it was a principle they fought for, as much as our ancestors, and not to avoid a three-penny tax on their tea; and the results of this battle will be as important and memorable to those whom it concerns as those of the battle of Bunker Hill, at least.

Although Thoreau was an observer of nature, he looked for a human lesson, a homily in each act of brute creatures. This anthropomorphizing compromises him as a pure scientist and at the same times permits him to embed each observation in the mainstream of human culture. If we have difficulty in categorizing Thoreau, it is only because his originality defies the usual pigeonholing.

The last chapter of *Walden* begins with a somewhat caustic attack on the adventurers of his century. As an explorer of the inner space of his own mind, Thoreau seems curiously antipathetic to those young men, like Darwin and Melville, whose explorations involved traveling far beyond their backyards into the far corners of the world. And so he writes: "Nay, be a Columbus to whole new continents and worlds within you, opening new channels, not of trade, but of thought." Yet Walden was to Thoreau what the *Beagle* was to Darwin or the whaling ship to Melville. They provided experiences that set directions for entire lives. There is something puzzling about that tirade against voyagers to far lands; it has a kind of sour grapes attitude about it that gives us a glimpse of the darker side of Thoreau's personality. He is viewing mankind by his own idiosyncratic standards, a very difficult judgment indeed.

But just when the man of *Walden* is beginning to disturb

us the most with his lack of understanding of others, he does an about-face and writes: "Let every one mind his own business, and endeavor to be what he was made. . . . If a man does not keep pace with his companions, perhaps it is because he hears a different drummer." Henry David Thoreau certainly heard a different drummer; sometimes the cadence took him into a study of science, sometimes into philosophy and ethics; always it moved him toward "truth": "No face which we can give to a matter will stead us so well at last as the truth. This alone wears well."

Although *Walden* is not usually thought of as a study of science, the pages of Thoreau's classic are alive with birds, insects, fish, trees, and shrubs. They were his neighbors for the time he lived in his handcrafted cabin in the woods. *Walden* also reveals its author's wide knowledge of mid-nineteenth-century natural philosophy. He was both a student of the universe and a careful observer of the microcosm he chose to make his personal world.

Humanizing
Scientists

Adlai Stevenson once began a speech in Cambridge, Massachusetts, with the words "I'm delighted to be here tonight, where I understand that at M.I.T. they are trying to humanize the scientists and at Harvard they are trying to simonize the humanists." In the interest of trying to humanize the public image of science, which often looks so cold and formidable, I would like to deal with an example from the genre of oral literature—those countless stories that researchers tell about each other at lunch or over coffee. I shall, mea culpa, recall an anecdote in which I was a participant.

Back in 1971, Huey Huang, who was then a postdoctoral fellow, and I were engaged in a study of mathematical methods of detecting ecological instabilities. We were trying to find early warnings given by systems about to crash, such as a lake on the verge of becoming eutrophic or a forest about to be totally defoliated by insects. The problem took us into some unfamiliar equations that we were unable to solve or even to make much progress with. In desperation we decided to seek out Yale's resident scientific genius, Nobel laureate Lars Onsager. He had founded the theory of irreversible thermodynamics in 1931 and over the years had made notable contributions to various fields of physics and chemistry. The term "desperation" is used because we stood in awe of this man's extraordinary accomplishments and were reluctant to intrude upon his time. But we also wanted to make some progress on the problem at hand. The research one is working on always assumes enormous importance in one's personal assessment of it. I remember a Ph.D. qualifying exam in which the professor asked, "What is the most important problem in modern biology?" The candidate turned to his questioner and replied, "What are you working on now, sir?" And so it goes.

Upon entering Onsager's office, we presented the equations to the guru, who made a few perfunctory comments. Then, in almost Zen-like fashion, he began to speak about something entirely different, the anomalous electrical properties of ordinary ice. Most solid electrical conductors, such as wires, carry electrical charge in the motion of negatively charged electrons moving toward the positive electrode. Ice is an electrical conductor that works by the jumping of positively charged protons from oxygen atom to oxygen atom toward the negative electrode. Onsager, who had worked out the basic theory of this phenomenon, proceeded to spend the next hour and a quarter explaining it to two very confused, unprepared scientists. We listened in due respect, thanked the professor, and left.

It is not an unusual story. One goes to a master with a question. That individual is so highly focused that he responds by discussing the area of reality within his penetrating gaze at the moment. The supplicant has no choice but to accept the wisdom being dispensed and go away in gratitude.

In the following years I saw Onsager only occasionally, and we discussed neither protons nor ecology, rather questions associated with the origin of life. We had a common interest in the importance of prebiotic membranes in directing the first protocells along the way toward biochemical activity.

The story would end here but for something that happened five years after the lecture on ice. I was then working on an entirely different problem—the mechanism by which energy from the oxidation of sugar is converted into a useful form for running cellular reactions, contracting muscles, and otherwise doing the necessary work to keep an organism alive. Peter Mitchell, a British biochemist, had proposed that the energy conversion was associated with the flow of protons across the membrane. There was a major dispute going on between Mitchell and others, who held a series of diverse views on the nature of the energy storage mechanism.

The energy conversions in question take place in mitochondria, tiny cell organelles highly specialized to process energy and package it in biologically useful forms. Our laboratory was studying the phenomenon in cauliflower mitochondria. Part of the investigation was a detailed thermodynamic analysis of energy changes. A focus on the mechanism by which the protons actually move from one side of the membrane to the other developed. Somehow in thinking about that problem, I dredged up from my memory bits of the earlier conversation with Lars Onsager. He had told me then what I now needed to know to understand transmembrane proton transport.

With those clues in mind, I read several of the professor's papers on the subject and wrote a brief letter to the editor of *Nature* postulating transmembrane proton conductance by a solid-state icelike device. The letter was rejected, but my copy to Onsager met a different fate. I had mailed it to his summer home in New Hampshire and heard nothing as the months passed.

On October 6, 1976, I was walking through the center of the Yale campus and noted the flag at half-mast. The card on the nameplate at the base of the flagpole read "Lars Onsager." A week later his reply to my letter arrived. He must have mailed it a few days before his death. It had been so long in coming because he had sent the parcel containing reprints by third-class mail.

The note stated:

Dear Harold:

Thanks for your letter. I am just back in Miami, where I keep most of my reprints. You will find the figure from the 1966 Boulder [lecture] reproduced on p. 13 of the Nobel lecture. The papers on protonics [ice] deal with important sources of quantitative background information from inorganic systems, and I thought you might be interested in "Early Days" [a paper on the origin of life]. I have promised a lecture on the "Early Days" at Yale on Nov. 9, and hope to see you on that occasion.

Cordially,

Lars

When the appointed day of his planned visit arrived, I paid tribute to Professor Onsager's memory by working diligently on the problem of membrane proton conductance—and wondering about the meaning of the whole series of events. There are those who would have seen evidence of the paranormal in the great scientist's seeming precognition of what I would need to know and his posthumous note commenting on it. Lars Onsager would not have been among them, for he was always given to the normal and the experimentally testable in his search for explanations. And I agree there is nothing paranormal in his lecturing to me, seemingly at random, about what I would need five years later. After all, he simply knew much better than I what was going to be important. After Onsager's death I sought out his former student and collaborator on the ice problem, John Nagle, professor of physics at Carnegie-Mellon University, and we began detailed work on the theory of transmembrane conductance.

There is such a thing as genius in science, and we can usually recognize it. It is extremely rare, involving a few individuals per generation. Most of us stand on the shoulders of such giants as Lars Onsager and try to peer ahead. We also establish the milieu wherein specially gifted savants can function. Additionally, the scientific community is involved in working out the consequences of ideas put forth by these individuals. Thomas Kuhn has explained to us how all this works in his book *The Structure of Scientific Revolutions*. Science is, after all, a highly social activity. And that is why we sit around telling stories about each other.

Sail On

I n one of those rare coincidences that so enliven the everyday scene, I found myself perusing Samuel Eliot Morison's Columbus biography *Admiral of the Ocean Sea* at the same time my wife was reading the more contemporary *Airborne*, a sentimental journey by William F. Buckley, Jr. On exchanging books one evening, realization dawned that we were both following Atlantic crossings under sail. The two trips were in different directions. With surprising spontaneity we invented "Sail On," a fascinating parlor game in which one compares Bill Buckley with Chris Columbus. I wish to share that game with those readers possessing a deep sense of history. As a caveat, it must be noted that the rules of the game disallow frivolous comparisons like those of obviously disgruntled liberals who claim that Buckley and Columbus have two of the finest fifteenth-century minds on record. We are interested in the more substantial stuff that scholarship is made of.

To keep the formal records straight, Columbus left the Canary Islands on September 6, 1492, and arrived at San Salvador island in the Caribbean Sea on the 12th of October, 1492. Buckley's voyage from Bermuda to the Azores took eleven days and eight hours sometime shortly after June 6, 1975 (William is less precise than Christopher about the exact dates; actually, he is less precise about almost all factual material, as contrasted to psychological or emotional aspects of the voyage). Buckley steered a course that was closer to Columbus's return voyage than to the outbound one. Both sailors knew the best latitude to sail from west to east. Columbus also stopped at the Azores, but only inadvertently, after being blown there by a storm. However, it is the maiden crossings of the two distinguished navigators that are the primary focus of the game Sail On.

A curious, indeed an almost mystical, resemblance seems
to exist in the physical appearance, facial features, and
certainly personality traits of the two men. This can most
impartially be seen if I record some comments made about
Christopher Columbus by his contemporaries. Since Buckley is a well-known public figure, the comparison and conclusions can then be left to each reader. The comments
from the late fifteenth and early sixteenth centuries are as
follows: "taller than average and strongly limbed"; "a nice
Latinist"; "gracious when he wished to be, irascible when
annoyed"; "long visaged with cheeks somewhat high . . .
an aquiline nose and his eyes were light in color—his com-
plexion too was light but kindling to a vivid red"; "face
long and giving an air of authority"; "eloquent and boast-
ful in his negotiations"; "in matters of the Christian religion, without doubt he was a Catholic and of great
devotion." Finally, to bring things up to date, we quote the
modern biographer Samuel Eliot Morison: "Columbus was
a man with a mission and such men are apt to be unreason-
able and even disagreeable to those who cannot see the
mission."

One of the crew on Buckley's ship was his son, Christopher, a man bearing the same name as the discoverer of
America. I do not know how the Buckleys came to name
their child or whether reference to the great navigator was
a matter of issue. Surely the Buckleys were aware of the
legend of St. Christopher that was known to Domenico and
Susanna Columbus when these two Genoese had their son
baptized in 1451. Buckley is clearly a Columbus fan, and in
Airborne he refers to the great admiral more than once.
Aspects of the Chris and Bill story may have roots in Buckley's early acquaintance with tales of Columbus's exploits.
In matters of descendants, Buckley and Columbus each had
a single legitimate son, and Columbus managed one other
male child by a less formal arrangement.

The 1975 trip was fifteen years in the planning, according to Buckley, who reports saying to Peter Starr about

fifteen years ago, "Let's face it. Someday we'll have to sail across the Atlantic." Planning for the 1492 voyage appears to have begun in 1477, after Columbus returned to Lisbon. At least we know of his correspondence relating to the subject with a Florentine geographer, Toscanelli. These letters may have begun as early as the late 1470s. In any case, it seems reasonable to assume that both voyages were approximately fifteen years on the drawing board. Not that Columbus needed such a long period for planning; in 1484 he tried unsuccessfully to get the backing of the king of Portugal for his expedition. Columbus had plenty of time and needed money, while Buckley had plenty of money and needed time. Of course, Buckley was nine years older than Columbus at their respective sailing dates, so he had almost a decade more to acquire the gold that appears to have been precious to both of them.

In the matter of comparing vessels for the two voyages, we lack a great deal of information about the flagship, *Santa Maria*. It may be presumed to have been slightly larger than the *Niña*, which Morison estimated at 70 feet overall with a beam at 25 feet. The *Cyrano*'s dimensions are more accurately known as 60 feet length overall and 17½ beam. Aside from the relative size and common use of sail power, the similarities cease.

Cyrano had auxiliary power in the form of a diesel engine; *Santa Maria* relied on oars of ash when other than wind power was needed. *Cyrano* carried an upright piano, nonoperating loran, nonoperating radar, automatic pilot, a calculator, original oil paintings, refrigerator, and air conditioning; *Santa Maria* did not. We know that Buckley has very strong thoughts on air conditioning, while historians appear to have totally overlooked Columbus's feeling on this issue.

Forty men sailed on the *Santa Maria* as it left the Canaries, all in the employ of Ferdinand and Isabella. The living arrangements were primitive, with no bunks being provided for the crew and only slightly better arrangements for the officers. The admiral, of course, had a cabin of his own.

The log of the *Cyrano* lists nine names, with well-defined class distinctions. First there was the captain (William Buckley), who, following the tradition of the sea, was a law unto himself (of mutinies we will comment later). Then there were five family and friends, who slept in private cabins. The rest of the crew consisted of three hired individuals, identified as the captain (the presence of two captains isn't explained), the cook, and the Argentinian steward. The employees seemed to move their sleeping arrangements according to weather and sea conditions. Clearly, the *Santa Maria* was a crowded, uncomfortable ship compared with the luxurious *Cyrano*.

Although both navigators used dead reckoning, Buckley relied heavily on celestial navigation, particularly sun sights. Celestial navigation was hardly developed in Columbus's time, and his few star sights were often severely in error because he had chosen the wrong star. Buckley had the more difficult task since he knew where he was planning to go, whereas for the early explorer any landfall would do. Of course, Columbus also had to worry about getting the ship back to Spain, which was not a trivial matter.

Of mutinies on board we have only veiled references. A few days before land was sighted, the crews of the *Niña*, *Pinta*, and *Santa Maria* were getting pretty upset about being so far from Spain, and the commander seems to have had to promise them a limit to the length of the outward voyage if no land was sighted. The captain of the *Pinta* went off on his own from time to time regardless of the orders from his admiral. The mutinies on the *Cyrano* seem to be of a strange psychological variety and show up in such bizarre log entries as "My ass, Buckley—DTM" and "Screw you—CTB." Both great men appear to have lacked the full respect and appreciation of their shipmates.

We could, of course, continue this round of Sail On for

many instructive pages, but I fear too much information would rob my readers of the fun of playing the game on their own. Suffice it to say that both ships completed their Atlantic crossings and arrived at islands where unsuspecting natives were left bewildered and amazed by the behavior of the captains and their crews.

Waving the Requirements

A while back, a student who was about to graduate from college announced plans to crew on an around-the-world sail. While few of today's professors would be likely to recommend such a voyage as the best possible way to begin graduate training, thoughts came to mind suggesting that in the early part of the last century a long trip aboard a sailing vessel seems to have been optimum graduate schooling. In some cases, it was also the best undergraduate education.

Certainly, in the decade from 1831 to 1841, a period at sea was highly productive for creative scientific endeavor. The two most far-reaching concepts of natural philosophy of the entire nineteenth century were conceived by young men as a result of observations and thoughts gathered on long sea voyages. Those scientific advances were the conservation of energy as formulated by Julius Mayer and the theory of evolution as set forth by Charles Darwin.

On February 22, 1840, twenty-six-year-old Mayer, newly graduated physician, set forth from Rotterdam as ship's doctor on the sailing frigate *Java*, bound for the East Indies. He had no research training, nor had he shown any propensity for scientific work. Fortunately, few clinical problems arose, and Mayer had time to read, study, and think. Apparently, he was something of a loner and engaged in little social conversation with officers and crew.

Like any good mid-nineteenth-century physician, he occasionally let blood as a therapeutic measure. Upon first carrying out this procedure in the tropics, he noticed that the blood was much redder than he was accustomed to seeing, and he initially thought an artery had been struck rather than a vein. Repeated observations on the redness of the return venous blood in the tropics compared with its much bluer color in cold north European climates set him to thinking. Mayer was, of course, making spectrochemical observations on the oxygen content of the blood, and he rightly concluded that venous blood was more oxygenated in the tropics. He postulated that this was the result of less oxygen being required to maintain body heat in warm environments.

The chain of biophysical reasoning led Mayer to a general formulation of the conservation of energy. He apparently intuited the right ideas even though his knowledge of physics was at that time quite inadequate. Upon returning to Germany to a general practice of medicine, Mayer wrote his seminal essay "On the Quantitative and Qualitative Determination of Forms of Energy."

Mayer's ideas were not immediately accepted both because he was outside the research establishment and because he had an often idiosyncratic view of physics. The importance of energy conservation was soon recognized through the work of others, and it was twenty years before the former ship's physician received full recognition for his groundbreaking work. In looking over Mayer's career, what seems exceptional is that his truly remarkable conceptual breakthrough was made in the virtually complete isolation of his cabin on the *Java*.

Like Mayer, Darwin was essentially an amateur at science when he embarked on his historic voyage. Darwin set sail from Plymouth, England, on December 27, 1831, as naturalist on the Royal Navy's *Beagle*, but he had spent his undergraduate days at Cambridge preparing for a career as a country parson. As the ship pulled away from the English coastline, Darwin began a five-year around-the-world adventure, during which he spent much of his time "unspeakably miserable from sea-sickness." The *Beagle* stopped at many ports of call, and Darwin, always eager to disembark, gathered botanical and zoological specimens at each land-ing. He made careful observations on geological formations as well as on flora and fauna, and he was the first to understand the growth of coral reefs.

Between landfalls there was a great deal of time to read, to think, and to study his newly collected specimens. By the time the *Beagle* stopped at the Galápagos Islands in late 1835, Darwin was an experienced, if self-taught, naturalist, and he had an opportunity to study and collect the unusual biota indigenous to that isolated oceanic ecosystem. From the examination of these plants and animals, Darwin apparently first glimpsed the idea of evolutionary descent.

Some twenty-four years later, after patiently collecting and sifting large quantities of data, Darwin published *The Origin of Species by Means of Natural Selection.* That calm yet revolutionary document has forever after structured biological thinking. In an analogous way, the theory of conservation of energy has dominated thinking in physics.

Mayer and Darwin both chose graduate training that seems strange by today's standards. Even in 1831, Darwin's father required great persuasion to allow his son to undertake the voyage on the *Beagle.* Both naturalist and ship's physician spent extended time in intellectual isolation and in that solitude were able to penetrate deep issues and arrive at truly innovative thoughts. It all seems so strange in today's academic environment.

Nor were Mayer and Darwin alone. Richard Henry Dana trained at sea as part of his undergraduate curriculum. In 1834, when measles weakened his eyesight, he left college and shipped out as an ordinary seaman. He sailed to California, was involved in gathering hides ashore, and returned in 1836 to reenter Harvard. In 1840 he published *Two Years Before the Mast.* The sensitive college student was impressed by the abuses endured by the common sailor. This resulted in a sympathy for the oppressed which lasted his entire subsequent career. This viewpoint was demonstrated in the free legal aid he gave to escapees captured under the Fugitive Slave Law.

For Herman Melville, whose family's financial position ended his formal education, a life at sea provided a substitute. He wrote, "A whale-ship was my Yale College and my Harvard." The author of one of the great novels of all time was largely self-educated, much of the formative thinking taking place on the whaler *Acushnet* (sailing from New Bedford in 1841), the whaler *Lucy Ann*, and the frigate *United States*.

Another author largely self-educated at sea was Joseph Conrad, who began sailing in 1874 at the age of seventeen. He spent the next twenty years as apprentice, steward, passenger, seaman, second mate, first mate, and captain. He sailed over much of the world and then settled down to become a highly regarded novelist and short story writer. Remarkably, although all his writings are in English, Polish was his native language.

John Masefield, the fifteenth poet laureate of England, substituted for college an apprenticeship on a windjammer that sailed from England around Cape Horn. He left the sea after that long voyage, yet the experience powerfully influenced his poetry, much of which deals with the lives, experiences, and emotions of sailors. Like Darwin, Masefield is said to have suffered from seasickness.

A pattern emerges. Gifted young men of the nineteenth century who did not fit the mold for one reason or another took to the sea. In this totally different environment, some of these adventurers grew to achieve a certain greatness in their fields of interest. The voyages were the catalytic experiences that set the individuals off on their careers. History does not, of course, record the unsuccessful ones, except in certain special cases like Masefield's poignant poem *Dauber*.

And what happens to today's gifted young people who do not fit the mold? I truly don't know, but I'm happy to say Godspeed to any graduating senior who tells me of a planned adventure outside of the academic mainstream. That experience may change his or her life. In certain exceptional cases, it may even change the world.

Dollars
and Sense

The Zen of Speculation

I picked up the phone and recognized the always very upbeat voice of my broker: "Hello, Harold, there's a stock that our analysts are very positive about, and I thought you might be interested."

Once again I realized that Smith Barney was making money the old-fashioned way—hustling—and I was prepared: "Before you tell me about the stock, I want to ask you a question."

"Okay, go ahead."

"Have you ever heard of Louis Bachelier?"

"I'm not sure. Does he write for the *Wall Street Journal*, or is he the greenmailer involved in the takeover attempt at International Papier-Mâché?"

"Neither. The Louis Bachelier I'm talking about was an economist who graduated from the Academy of Paris in 1900 and had his thesis published in the *Scientific Annals* of the famous French École Normale."

"Oh, is his manuscript being sold by Sotheby's? Are you going in on a syndicate?"

"No, no, that's not the problem I want to talk about. It's Bachelier's doctoral thesis I have to discuss with you."

"I get it," said the stock salesman. "In those old books you're always reading, you've found a way of making money, discovered by some Frenchman."

"Not exactly," I said. "Louis Bachelier was a mathematical economist who studied fluctuations in prices on the Paris stock exchange. He was a remarkable man who developed a complicated set of formulas to study market changes." And before he could protest, I read to him this excerpt from Bachelier:

The influences which determine fluctuations on the
Exchange are innumerable; past, present, and even discounted
future events are reflected in market price, but often show
no apparent relation to price changes.

. . . Contradictory opinions concerning these changes
diverge so much that at the same instant buyers believe in a
price increase and sellers in a price decrease.

The calculus of probabilities, doubtless, could never be
applied to fluctuations in security quotations, and the
dynamics of the Exchange will never be an exact science.

But it is possible to study mathematically the static state
of the market at a given instant, i.e., to establish the law of
probability of price changes consistent with the market at
that instant.

There was a profound silence on the other end of the
line.

"Are you still there, Tom?" I asked.

"Yeah, yeah, but now I'd like to tell you about this new
issue. Do you know anything about the monoclonal antibody industry?"

"Wait, I'm not through talking about Bachelier. He went
on to study the fluctuation of prices on the Paris Exchange,
and he developed the very same mathematical formula that
Albert Einstein used to study the movement of microscopic
particles, the so-called Brownian motion. Only Bachelier
worked out his theory a full five years ahead of the famous
Einstein."

"That's nice; but there's this new company in Castro-
ville, California, called Gene-choke Inc. that makes monoclonal antibodies to artichoke leaf blight bacteria. It's going
to revolutionize the whole industry. I'm going to buy this
one for my own portfolio."

"Wait, I'm not through yet," I interjected. "I've got
more to tell you about Einstein. His equation for Brownian
motion describes chaos, complete randomness; it's impossible to predict which way a particle will go. In the same
way, Bachelier showed it's impossible to predict which way

the price of a stock will go. That too is in the domain of chaos, just like the particle motion studied by Einstein. The bottom line, Tom, from Bachelier's studies is that over the long term, without illegal insider knowledge you can't beat the market averages. You don't by any chance have any illicit inside information?''

Again I sensed a pregnant silence before my ever optimistic broker began. ''The president of Gene-choke is Dr. Peter Plentiful. You may have heard of him. He made a bundle with Gene-uflect ('We bow to no one in our knowledge of biotechnology') before they were bought out. This is really a great opportunity to get in on the ground floor. What do you think?''

I continued. ''You're not listening. If the market is mathematically chaotic, then buying Gene-choke is not too wise. I have just as much chance of losing money as of making money.''

''But we have specialists.''

''If the market is chaotic, there are no such things as specialists. It's like having specialists in where a roulette wheel is going to stop. But nevertheless, Bachelier is going to help me in the stock market.''

''How is that?''

''By convincing me to get out.''

Smith Barney's man, ever true to his craft, started to plow ahead again. ''There is some more about Gene-choke Inc. that you should know. They have a licensing agreement with the Plant Pathology Department at San Jose State, and 20 percent of their stock is owned by the Swiss firm of Biohype.''

I interrupted. ''No, I'm not interested. My new investment strategy is the Zen of Bachelier, which is the same as the Tao of Jones. Just as the Tao orders the chaos of the universe according to early Chinese philosophy, I need to overcome the chaos of the stock market in my investments.''

''How do you do that?''

"By getting out of common stocks and investing in bond funds, certificates of deposit—anything that by conformity to unassertive action and simplicity guarantees me the tranquility of the Tao."

"Well, let me tell you just one more thing about Genechoke Inc., and this is supposed to be a secret, but I'm confident it's correct. They have a chemist who can extract from artichoke hearts a new tranquilizer that is supposed to be better than Valium."

"But if I listen to Bachelier and follow the Tao of Jones, I won't need a tranquilizer."

"Well, I tell you what, Harold. Why don't I just send you a prospectus, and you can look it over and decide for yourself."

"You do that, Tom; just send me the prospectus."

The Entrepreneurs

N ot long ago I visited a zoologist in the western United States and heard reports that the extraordinary mountain runoffs this year were pouring so much water into Great Salt Lake that the decreased salinity was endangering the brine shrimp. This is not a classic "endangered species" problem, since this creature is broadly distributed in North America. However, hearing about these curious little crustaceans revived some long-dormant memories.

This tale deals with three serious young scientists launching forth in an effort to corner the world market in a most unlikely commodity. I will recount how the entrepreneurial instinct seized these eager scholars and propelled them on a whimsical adventure.

The time was 1955, and the place was a university research laboratory. The issue being debated dealt with biological information: since bacteria could be dried and frozen to temperatures near absolute zero and subsequently revived and induced to grow, did that not imply that cellular information was entirely structural in nature and that no continuity of process was necessary? The conclusion seemed to be a clear "yes," but, argued the skeptics, what about the far greater information in cells of more complex organisms?

The search for a more elaborate system to test led to some experts who reported on Artemia, or brine shrimp, animals whose eggs dry out every year in the mud flats of the Great Salt Lake. When the spring waters come, the lake level rises and the eggs are rehydrated and continue their developmental cycle, hatching out as tiny swimming brine shrimp nauplii. The eggs can still be revived after long periods of drying.

Shrimp eggs, each a small fraction of a millimeter in size,

were obtained and experiments begun. The dried eggs, which usually take the form of collapsed spheres, contain immature brine shrimp embryos. When placed in salt water they swell to spherical form in about an hour. A few hours later the shell splits open and the developing embryo emerges, still surrounded by a protective membrane. Some fifteen to twenty hours after being placed in brine, free-swimming nauplii break out of the membrane and take off to find food.

After repeating these experiments, one of the scientists sought the cooperation of the low-temperature group in the physics department. A batch of eggs was taken to within two degrees of absolute zero and held in that state for seven days. On rewarming, the eggs were immersed in salt water and showed the same hatch rate as control eggs kept at room temperature. The point was proved: fundamental biological information, even for something as complex as a crustacean embryo, is structural and requires only the right environment for structure to become function. The question of whether this applies to learned information is yet to be resolved.

It occurred to the scientist making these observations that Artemia had another possible use; they could be put in children's microscope sets. For when that gift is presented on Christmas morning, there are few living specimens to examine in most of the winter-bound United States. Watching shrimp hatch would be just the thing to increase the play and educational value of the microscope set. The young man consulted with two colleagues, one of whom was knowledgeable about the source of Artemia and the second of whom had a contact in the A. C. Gilbert Company, then one of the country's leading manufacturers of scientific toys. A partnership was immediately formed, and some time later a contract was entered into between the three scientists, "referred to hereinafter as the Sellers," and the A. C. Gilbert Company, "referred to hereinafter as the Buyer."

The contract called for the sellers to write an instruction manual and to refrain from dealing with other toy manufacturers. The sellers also had an exclusive two-year supply contract for brine shrimp eggs. By Christmas of 1956 the microscope sets did include Artemia eggs and The Gilbert Shrimp Manual, "another Gilbert Hall of Science product."

The idea took hold, and the following year brine shrimp eggs were widely found in scientific toys. Ads for "sea monkeys" began to appear in comic books, and Artemia took its place among children's toys and scientific teaching aids.

When the time came to renegotiate the supply contract, the three scientists conceived an idea bold in its simplicity and geared to the American dream. Why not corner the market in brine shrimp eggs and thereby control the price for all users of this commodity?

Phone calls were made, orders were placed, and the freight began to roll from Ogden, Utah, to New Haven, Connecticut. In the basement of a house in suburban New Haven boxes began to pile up, each with four one-gallon paint cans full of dried brine shrimp eggs. Visions of sugar-plums and yachts and sports cars danced in the heads of the three "businessmen" as they presented the revised price schedule to corporate management. As days and weeks passed, however, their euphoria turned to concern. That concern deepened when the sellers found out that the buyer had found a new supplier.

What the scientists had ignored were two facts, one ecological and the other financial. *Artemia salina* is a very cosmopolitan species, and the toy company had found an independent supplier in Louisiana. The manufacturer, driven by the usual motive (finding the lowest price for a commodity), had displayed more ecological wisdom and financial know-how than the entrepreneurial biologists.

The sellers were now warehousers, their profits tied up in a basement full of paint cans brimming with Artemia eggs. It was truly a learning experience, and driven by economic necessity, the entrepreneurs set out to develop new markets. Fortunately for them, there was a totally different use for brine shrimp. Fanciers hatched out these tiny animals as food for their fish, so pet stores sold Artemia in appreciable quantities. The basement paint cans were unloaded at cost, and the scientists retired from their business careers.

This story seems to call for a moral or a series of morals, particularly in these days when scientists are getting more and more involved in commercial ventures. However, most good morals have been exhausted by Aesop, so perhaps an epilogue might be more to the point.

The A. C. Gilbert Company has gone out of business for reasons unrelated to this venture. The scientists continued on in careers in biology and medicine. The most lasting effect of the whole episode has been the continuing use of brine shrimp eggs in scientific toys. I keep meeting undergraduates who have had some experience with Artemia as part of their formal or informal education. And that is pleasing for, as you may have guessed, I was the scientist who inspired these chastening entrepreneurial careers.

Where Am I?

Most yachties I know have long since given up the angst of asking. "Who am I?" and have settled for the less philosophical but nonetheless tricky question, "Where am I?" The answer at sea is always numerical: latitude and longitude, the two coordinates necessary to locate unambiguously a point on a spherical surface—or a ship on the ocean. The method of determining position used by mariners is in principle quite simple, although in practice, taking a sextant reading and performing the calculations on a sea-tossed boat require an appreciable amount of skill and experience.

After 3,000 years of observing and recording, we are now able to predict at any time of day or night the point on the earth's surface directly below any celestial body—sun, moon, planet, or star. This information is all listed in a government publication called *The Nautical Almanac*. An observer wishing to find his or her location measures the angle between the heavenly body in question and the horizon. The angle establishes the distance between the observer and the point under the object sighted, which is found in the almanac—if one knows the exact time of sighting. A series of at least two such measurements separated in time will yield a point location. Since I've never been required to perform those operations at sea, I make the above statements with great certainty and confidence.

The preceding paragraphs represent, for me, very recently acquired knowledge, the result of a decision to learn something of navigation. In a number of conversations lately, students keep asking, "Say, where are you coming from?" and I thought it would be nice to present a precise answer. Also, I once had to use latitude and longitude to convince *National Geographic* of my mailing address.

Having perused a number of introductory texts on the subject of interest, I have uncovered a dirty little secret that I feel compelled to share, and frankly, I'm embarrassed. So if you just turn your eyes away, I'll whisper it: "Navigators use Ptolemaic astronomy." If that doesn't shock you as much as it did me, I'll have to review what we teach the young about the history of science.

In ancient days people looked up at the skies, observed the motions of the heavenly bodies, and naturally concluded that the earth stood still while the stars in their heavenly sphere turned around us. The sun, moon, and planets, each in its own celestial sphere, all circled about the earth. In the second century of the common era, the Alexandrian astronomer Ptolemy (Claudeus Ptolemaeus) formalized this theory into a tight mathematical system.

The geocentric theory stood for over 1,300 years, until 1543. That date marks the publication of *De revolutionibus orbium coelestium (On the Revolutions of Celestial Spheres)* by the Polish savant Mikolaj Kopernik (Copernicus). A lifetime of calculation of celestial motions convinced Copernicus that the simplest explanation of observed events was to assume that the sun was at the center of the universe and that the earth moved around it yearly and rotated, as well, on its own axis every twenty-four hours. That was truly one of the great advances in human thought, one that was vigorously opposed by the Church, as exemplified by the house imprisonment of Galileo for his support of the Copernican model. In any case, the doctrine of the central sun led to an expanded view of the universe, to the physics of Galileo and Newton, and to the modern age of science. The beginning of contemporary thought dates from the discovery of the earth's motion around the sun.

The heliocentric view of the universe has become universally known and accepted since 1600, and students of the physical sciences celebrate the victory of the Copernican doctrine over the Ptolemaic. That change of viewpoint is one of the prime case histories chosen by Thomas Kuhn to demonstrate a scientific revolution, or, in his words, "a particularly famous case of paradigm change." In addition to having scientific impact, Copernican theory transformed our philosophical view by removing us from a privileged position and making us less anthropocentric. In short, every schoolchild comes to know that the modern world began when we threw off the blinders of Ptolemaic astronomy and opened our eyes to the brilliant light of Copernicus' sun shining at the center of the solar system.

Thus, the notion of using a fixed earth with moving celestial bodies in order to navigate comes as a considerable surprise. It's as if someone told us to assume the flat-earth viewpoint or the phlogiston theory of fire. It seems fitting when ideologically conservative-minded mariners like Bill Buckley take this pre-1543 point of view toward navigation, but when liberal modern humanist sailors like Dr. Benjamin Spock adopt the same position, it is really quite another thing (see William F. Buckley, Jr., *Airborne: A Sentimental Journey*, p. 47, for a description of a brief encounter between yachtsmen Buckley and Spock).

After a detailed reading of some introductory works on celestial navigation. I think I can now explain the strange pre-1543 attitude taken by today's navigators. First, the task at hand is to establish one's position on the earth, not to establish one's position in the universe. Therefore, one's only interest in other celestial bodies is confined to those that lie over a given point on our planet's surface at any time. *The Nautical Almanac* provides locations in earth-centered latitude and longitude values and designates them as GPs (geographical positions). Any celestial body is sufficiently far from the earth that all the light rays striking the earth from that source are parallel. Therefore, the exact distance between earth and observed body becomes unimportant for navigational purposes.

Given the previous considerations, most navigators find it much easier to think of the celestial bodies as moving, which is of course what one naively observes. Sailors take measurements, work out their position, and get the right answer because *The Nautical Almanac* is independent of cosmological theories. From an abstract mathematical orientation, Ptolemaic and Copernican astronomy are just different perspectives on the same events. And we have the empirical evidence that sailors usually get where they are going.

Well, having understood this strange archaic usage, what moral can we draw? If you want to know who you are, you should adopt the Copernican doctrine and use the best and most modern view of our universe to move most advantageously from contemporary science to its metaphysical sequelae. If, on the other hand, you desire to determine your location, take any viewpoint that works. After all, wanting to know such practical information is, to say the least, a very anthropocentric activity. As I said at the beginning, yachties are much more interested in establishing where they are than in thinking about who they are.

Madame Pele

Call me a vulcanophile. While Ishmael may feel the call of the sea, I am drawn to the sights and sounds and brimstone odor of molten lava pouring from the ground. Watching this birth of new land is for me a religious experience, the alpha and omega of planetary death and resurrection all combined in a single event.

It is difficult to arrange one's vacations around such episodic events as lava flows, so great flexibility is necessary. It is New Year's Day, and I find myself on board a plane flying from Kahului, Maui, to Hilo, Hawaii. I have no automobile or room reservations. In short, in a Thoreauvian sense, I am free. But to be free today is either to be rich enough to buy one's way out of any snags or to be willing to take a backpack and live in any setting that nature provides. I fit neither category, being a victim of a modicum of middle-class respectability, the great enemy of freedom. My only chance for liberty is an occasional attempt at Emersonian self-reliance.

The first challenge comes quickly. There are no cars for rent on the island for the next five days save for a single Suzuki off-road vehicle. I opt for this transportation at an obscenely high rate and then proceed to Hilo to find one of the least expensive hotels in town, the old reliable Hukilau.

At dawn I'm off on the road to Hawaii Volcanoes National Park headquarters to check out sites for viewing the new, large, still active flow. There are three possibilities: Chain of Craters Road from the park to where the flow crosses the road, Chain of Craters Road from Hilo to where the flow crosses the road, or overhead.

The next stop is Volcanoes Helicopter, where the first news is that all flights are booked for the next two days.

My name is placed on various waiting lists, and I take off down Chain of Craters Road. The twenty-eight-mile ride from park headquarters is spectacular even without a live volcanic flow. At the beginning of the drive through a Metrosideros fern forest, I begin to take notice of my "wheels." At first I had had unkind thoughts about it because it was costing so much money, and I had overlooked the fact that it is a spanking new, bright-red, sassy vehicle with the model name of Samurai. There is no other traffic on the nicely paved Chain of Craters Road, and as the Samurai picks up speed I begin to experience the positive pleasure of driving, a feeling I have not had in many years of traveling the car-choked highways of the northeastern United States. Rolling along through the scrub grassland down by the sea, a tall-in-the-saddle machismo overcomes me and my red Samurai, and a burst of discipline is required to contain the speed of rider and ridden.

On the highway is a roadblock for cars, and farther along, a second barrier has a warning for pedestrians. Just a few yards ahead, lava covers the highway and one can see the flow from somewhere up in Kilauea extending down into the sea. The ground is covered with fresh black pahoehoe lava and the remains of burned trees. Starting near the roadblock one can hike to the shore and see the shiny black sand produced when the molten basalt hits the water and vitrifies.

A mile or so back from the barrier is a national park visitors' center, the Wahaula Heiau, where an ancient Hawaiian temple once stood. The information center is presided over by Auntie Lei. A month ago her house was destroyed by lava. She speaks of her loss in a calm, resigned manner, and our talk turns to Madame Pele, the volcano goddess. She is not a malevolent goddess, for Auntie Lei reports that although Pele takes property, she does not take lives, and that is what is important.

Although we vulcanophiles take modern geology and

vulcanology very seriously, we also find it difficult to ignore Madame Pele and other volcanic deities. I know these eruptions occur because the Pacific tectonic plate is rotating and this island is passing over a deep lithospheric hot spot, but somehow Madame Pele remains a vivid local personality.

I remember some years ago being at Omar the Tentmaker's in Honolulu to rent a backpack and sleeping bag for a hike up Mauna Loa. Omar and I had been discussing the trip, and as I was leaving he called me back. "One more thing," he said. "Madame Pele does not like it if you pee on hot lava." I have certainly never offended her that way, and the volcano goddess and I have been on good terms. More devout local followers of Madame Pele placate the goddess by throwing bottles of gin into the molten lava.

Auntie Lei and I have a good conversation. She shows me pictures of her house before, during, and after the lava flow. The "after" photo shows just black rock, for the flow has completely covered the burned remains. Auntie is a very stoic lady; she knows the island was built entirely of lava flows. It is not possible to love this island, which she does, and at the same time to hate the hot lava. At least Pele gives sufficient warning for the people to escape.

I drive back to the office of the helicopter service and find, to my delight, that there has been a cancellation on the three o'clock flight. The copter view of the volcano begins at the shoreline, the area I have just seen from the ground. The aerial view affords some sense of the substantial area covered by the latest flow. Flying back toward the vent, we can see areas where red molten lava still glows through. The vent itself is in a large caldera of lava mostly covered with a thin gray film of solidified rock, with large molten bubbles occasionally breaking through. Underground tubes carry the lava down the mountainside. The experts have no idea how long the flow will continue.

The short helicopter ride is over, and I remount Samurai to return to Hilo.

Early the next morning, it's off to Kalapana, where several houses have been destroyed by Pele. There are remains of metal roofs, and a truck and tractor are stalled axle deep in newly hardened rock. Other artifacts are firmly embedded in this black basalt. Near the lava line stands a house entirely unscathed, only 200 feet from one that has been totally demolished.

Most geologic events occur so slowly it takes a lifetime to notice any change; not so on the slopes of Kilauea. There, at times, one can watch the advancing river of lava continually changing the landscape. It is perhaps that lesson in the impermanence of all things that makes vulcanology so fascinating. While we are now all aware of the great tectonic processes by which the earth recycles its chemical stuff, they seem to be intellectual abstractions. To stand on hard rocks that were liquid only a few days before engenders a very tangible feel for the ever-changing nature of our planet. Then again, I may just love this place because the always unpredictable Madame Pele is such a vivid symbol of freedom.

Reference Books

I have always been a lover of libraries. Being surrounded by books gives me a sense of womblike security. Only occasionally have disturbing thoughts broken through the traditional quiet of the printed word. When I was a boy spending a Saturday afternoon in Adriance Memorial Library in Poughkeepsie, New York, I timed how long it took me to read a single page and carried out a calculation that convinced me that not in a lifetime would I be able to read all the books in that institution. That's the kind of shocking numerically imposed humility that comes hard to a twelve-year-old.

Although my library time is usually spent at large institutional book collections, occasional R and R brings me to the Lahaina Public Library in Hawaii. It is, of course, a small collection and at first I wondered if it would answer my reference needs. But a rapid inventory showed that the shelves contained some fine old friends: *Encyclopaedia Britannica, Handbook of Chemistry and Physics, Physicians' Desk Reference,* and *Encyclopedia of Philosophy.* That start provided a certain reassurance that no really desperate lack of information would occur. After all, one never knows when the boiling point of toluene, the philosophy of Benedict Spinoza, or the uses of scopolamine hydrobromide will become matters of some interest.

This recalls those wonderful discussions of which ten books you would want with you if you were to be alone on an uninhabited island. I'm sure of *Moby-Dick* and the *Handbook of Chemistry and Physics.* About the other eight I'm still undecided. I once heard a story of a man who had to winter over in northern Alaska with nothing to read but an introductory calculus textbook. He wasn't much of a generalist, but after that winter he could differentiate and integrate complicated functions with great facility.

My hours in the Lahaina Library have been enlivened by people-watching, viewing the heterogeneous collection of locals, tourists, and transients who pass through. I know of no other library where the reader at the next table may be barefoot, bikini-clad, or bedecked in some other strange garb. It is hard to tell the bums from the déclassé intellectuals. Of course, that is probably a long-standing problem—classifying the human flotsam and jetsam that have washed up on Maui's shores over the past 200 years.

One day in the library I met a young man who had just dropped out of a McGill University MBA program and was concentrating on nineteenth-century English romantic poetry. He alternated between reading and staring out the window at Lahaina Roads with its varied collection of boats riding at anchor. He was definitely into Shelley, Byron, and two-masted sailing vessels.

On another occasion I met a woman who was about to sail to the small island Palmyra, where she and her husband would be the only residents for some time. She was eagerly reading all the latest magazines, as if to store up a reservoir of knowledge of the contemporary world to take with her into isolation. Her reading was pursued in a very business-like way, reflecting the importance of salting away an adequate amount of contact with the rest of humanity.

My last journey to Lahaina took me back to my reading table next to the current magazines. It was a hot afternoon, but a stiff breeze off the water kept the building comfortable while mandating the use of books as paperweights. I was pondering a quaint and curious volume on solar energy when I sighted, in the corner of my eye, someone entering the room. He was of medium height and light weight, and long unkempt blond hair fell to his shoulders. He wore cut-off jeans and an obviously aged T-shirt, and on his feet were a pair of worn rubber thongs. With an intermediate growth of beard the image was hippie through and through.

Our young visitor walked up to the librarian and made a quiet inquiry. She pointed to the reference shelf and he walked over and pulled down one of my beloved books, the *Physicians' Desk Reference*. The incongruity of such a strong interest in pharmacology in this particular library user crossed my mind as I watched him thumb through the "Product Identification Section," the "actual size full-color reproductions" of all the tablets and capsules sold by the major drug manufacturers. He quickly located what he wanted and turned to the product name index. He examined a section of the main text, read one or two pages very quickly, shook his head, closed and shelved the volume, and exited. Thus ended the vignette.

I just sat and stared, wondering what bit of biochemical knowledge could have been so interesting to this young hippie in paradise. I speculated at length, and I'm sure each reader will have a favorite candidate for this library user's drug of the week.

When I returned to the dock and told my story to some more-streetwise people, they assured me that *PDR* was a well-known work in the drug subculture. I had stumbled across a use of it that I was unfamiliar with—but that, doubtless, was the result of my sheltered existence. New knowledge about one of my reference favorites took me back to the library, where I checked the foreword and reread the publisher's purpose: "Intended primarily for physicians, *PDR*'s purpose is to make available essential information on major pharmaceutical and diagnostic products.... In making this material available to the medical profession, the Publisher does not advocate the use of any product described herein." Over two thousand drugs are described without advocacy in this medical tome.

It's now time to think about whether *PDR* should go into the collection of ten books to be taken to my deserted island. The answer is no, of course, since the usefulness of the book depends on the backup of the pharmaceutical industry, which would be unavailable. On the other hand, if my man Friday on this particular isolated site happened to be a certain young hippie with plastic bags full of pills of various shapes and colors, I'd have to rethink the whole matter and perhaps finally add *PDR* to my hypothetical library. Friday might even have his own copy by then.

I Must Down to the Seas Again, [For] The Woods Are Lovely, Dark and Deep

Some years ago, I was associated with the publication of an anthology containing John Masefield's celebrated poem "Sea Fever." Shortly after the book was released, calls and letters came to the publisher complaining of a mistake in the first line, which was printed as

> *I must down to the seas again,*
> *to the lonely sea and the sky,*

The callers and writers were eager to inform those responsible that the poem was a lifelong favorite of theirs that properly begins:

> *I must go down to the seas again,*
> *to the lonely sea and the sky,*

These complaints sent me scurrying to the library, where a search through numerous poetry anthologies revealed both forms occurring, with a definite preponderance of the second reading, containing the insertion of "go." This is the kind of challenge that sends one off, however ill prepared, in a burst of literary scholarship.

When the opportunity occurred, I found myself at Yale's Beinecke Rare Book and Manuscript Library searching for the earliest printed version of the poem in question. After presenting myself to the librarian, undergoing the appropriate security clearance, and divesting myself of all pens, I entered the reading room, the sanctum sanctorum where the rare volumes are presented to scholars for perusal.

In about ten minutes a small, thin copy of a volume entitled *Salt-Water Ballads* was brought into the room and put down on the reading table. The publisher was Grant Richards, of 48 Leicester Square, London, and the date of publication was 1902. A feeling for the authenticity of the source was considerably enhanced upon my finding on the first page the handwritten words "from John Masefield January 1903," accompanied by a small cartoon of a dancing figure bedecked in bell-bottom trousers. And on pages 59 and 60 there appeared the poem "Sea Fever," beginning, "I must down. . . . "

Elation seemed out of place in the somber setting of the rare book room. There was a sudden feeling for the emotions of the people who engage in that kind of scholarship. It's not only we scientists who have our "eurekas." There was, however, no surprise at the outcome, for I had made a bet with myself that the first version was the correct one.

The question then raised was, Where did the "go" come from? I do not have a definite answer, but circumstantial evidence suggests that at some time between 1902 and 1918 an anthology editor, compiler, or typesetter introduced the additional word and that that edition was the source from which many succeeding anthologists selected their poem rather than going back to the original source. One wonders how the permissions problem was handled. The practice of not seeking the original version is very understandable from a pragmatic point of view. We rarely travel to rare book collections to seek the canonical version of great writings. The real guilt lies with the person who first intruded on the poet's words with a gratuitous addition.

My experience with a single poem of Masefield's recalls a fascinating article by Donald Hall entitled "Robert Frost Corrupted" (*The Atlantic Monthly*, March 1982). Hall takes to task Edward Connery Lathem, editor of *The Poetry of Robert Frost*, for making 1,117 gratuitous emendations in Frost's poetry, an average of 3.4 per poem. Hall writes, "Lathem has removed commas, added commas, removed hyphens, added hyphens, made words compound, added question marks, and altered dashes." As an example, the line

The woods are lovely, dark and deep.

becomes

The woods are lovely, dark, and deep.

The added comma alters both meaning and sound; what else is left to poetry?

Hall's article is a seething indictment of Lathem, and I endorse every word of it. I can get a trifle annoyed at someone's trifling with Masefield, but anyone who touches a comma of Robert Frost's incites me to "Fire and Ice." I believe that Hall's essay should be made must-reading for everyone in publishing, and if *The Atlantic Monthly* gives me permission, I'm prepared to send a copy to every bona fide editor who requests it (even those at *Hospital Practice*). How's that for commitment?

The reason I am so irate over unapproved changes in a published text is that I believe that such editorial acts destroy history. It is the white-collar equivalent of book burning. Sequential application of the practice reminds me of a childhood game. In a roomful of players one person would whisper a phrase into the ear of the next, who would similarly pass it on to neighbor on the other side. When the final version was recited, it was almost always so transmogrified that it bore little relation to the original. So, through the ages, have certain editors, proofreaders, and emendators robbed us of our heritage by assuming that they knew better than the original author how a certain piece should read. With literature as individualistic and idiosyncratic as poetry, the full folly of their position becomes manifest. The sin of editorial pride is clearly a serious matter.

I now for the first time fully appreciate the historical tradition by which the Pentateuch has been transmitted in traditional Judaism. In setting forth the scribe's duty, *The Jewish Encyclopedia* notes:

> He is obliged to have before him a correct copy; he may not write even a single word from memory; and he must pronounce every word before writing it. Every letter must have space around it and must be so formed that an ordinary schoolboy can distinguish it from similar letters. . . .

Much of modern scholarship in history consists of trying to reconstruct the past from altered sources. That generates Ph.D. theses, but it nevertheless leaves us uncertain about even relatively recent happenings (in 1902, for example). We cannot undo the existing errors, but we can commit ourselves to higher standards of editorial responsibility. Well, enough, for now:

> *I must down to the seas again,*
> *[For] The Woods are lovely, dark and deep.*

Stand-up Conic

There it was, right in *The New York Times*. They were accusing my friend Tom Duffy, Yale band director, of being "hyperbolic." Since I had always considered him as being pretty straight, I pondered "all the news that's fit to print." Of course, "hyperbole" was a familiar term going back to the rhetorical discussions of Isocrates and Aristotle. But seeing the word in its adjectival form was jarring, because "hyperbolic" was familiar to me from mathematics (hyperbolic coordinate system, equations, functions, etc.) rather than from more literary uses. In any case, if the editors of the *Times* had used the traditional "hyperbolical," no philosophic(al) confusion would have resulted between the mathematic(al) and rhetoric(al) applications.

The challenge of sorting out all of those semantic problems was irresistible. Indeed, both "hyperbole" and "hyperbola" come from the Greek ὑπερβολή, a combination of ὑπερ (over) + βάλλω (to throw), to throw over, to exceed. In analytic geometry a hyperbola is a curve with two arms extending to infinity, which is not a bad metaphor for a hyperbolical orator or, in music, for an enthusiastic conductor. The musical sense was not intended in the *Times* article.

The oratorical and mathematical usage immediately suggested "circular," from the Greek κίρκος, or ring. The term applies equally to the mathematics of a perfectly round object or to a fallacious mode of reasoning.

While we were sitting around and discussing that interesting parallelism, someone suggested the case of "parabolic." Here the derivation is exactly of the same form as "hyperbolic." the Greek παρά (beside) plus βολή (casting) led to casting or placing beside, as in the comparisons of *parabolē* or as in the two equal arms of the parabole lying symmetrically on either side of the axis.

The oratorical *parabolē* is much more familiar to us in the form of parable, the metaphorical discourse so familiar in the stories of the New Testament. And considering the moral tale Tom had to tell, i.e., his side of the story, I think the *Times* would have done better to refer to him as "parabolic." Under no circumstances was he circular.

All of the discussion began to focus on what the circle, hyperbola, and parabola have in common. They are, without exception, conic sections, curves formed by the intersection of a cone and a plane. The form of the curve depends on the angle of intersection. There are, of course, only four such curves, the fourth being the ellipse. That realization sent us scurrying to the dictionary, and we were not disappointed. The adjective "elliptical" can come from the geometric ellipse, or oval, or from the oratorical ellipsis, the omission of one or more words in a sentence. Both trace back to the Greek ἐλλείπω, to leave out or to come short. In any case, Duffy could never be accused of being elliptical, for he is definitely not known ever to be short of words.

The discussion mandated a trip to the library, where I followed the path of Aristotle by Euclid to Apollonius of Perga. The last savant in that list, "The Great Geometer," lived from 262 to 190 B.C. in a region that is now part of Turkey. In the classic work *Conics* he is supposed to have introduced the mathematical terms "parabola," "ellipse," and "hyperbola." Apollonius' work was an extension of a lost book by Euclid, who apparently did not use the finally adopted nomenclature for conic sections.

The rhetorical origins of "parabola," "hyperbola," and "ellipse" are less well defined, but they predate Apollonius' work by at least two hundred years. Aristotle, in *Rhetoric* (Book II, Chapter 20), speaks of "the illustrative parallel," or *parabolē*, and gives us a case in point in an argument by Socrates. Fables by Stesichorus and Aesop are a related construction discussed in Book II. They are both included in the category of "examples." Lack of

knowledge of Greek prevents me from tracing these matters further, but doubtless a doctorate in classics could
emerge from a study of the rhetorical roots of the terminology on conic sections.

At this point I am beginning to hear echoes of F.S.C.
Northrop's course on the philosophy of science. I can't remember the exact date, but it was some time after Apollonius' work on conics and before Alan Turing's article
"Computing Machinery and Intelligence." Professor Northrop stressed the relations between science, philosophy,
and religion, particularly in the context of Greek culture.
He was especially impressed with Euclidean geometry as a
paradigm of a mathematical viewpoint that had a profound
effect on post-Euclidean culture. Within the framework of
his thinking, the parallelism between mathematics and
rhetoric would have been expected rather than surprising.
I was being treated to an understanding of something I had
been taught but had not comprehended in my undergraduate days.

During my etymological researches I stumbled upon another word from the same root as parabola. Admittedly
unrelated to our present discussion, it has such inherent
nobility that I want to share it. In the early Roman Codes
occurs the Latin form *parabolani*, persons who risk their
lives as sick-nurses. I immediately envision Father Damien
living with the lepers on the island of Molokai and eventually dying of the disease. But I am sure everyone has a
parable of a *parabolanus*, and it seems a pity that English
lacks a single word to express the concept quite so eloquently as the Latin.

Well, we have gone a long way from Duffy's alleged hyperbole to Apollonius to Euclid to Aristotle to the Justinian
Code to the passion of Father Damien. I sit at my desk
seeking a moral. None is forthcoming, but I do want to
thank the editors of *The New York Times* for sending me
off on such a fascinating excursion into the Greek foundations of our philosophical and rhetorical concepts.

Metaphysical Musings

Optimism as a
Moral Imperative

T here exists in the Western world a long tradition of intellectual pessimism that is regarded as fashionable by the cognescenti and their followers. Although the movement has clear Continental roots, it has spread to England and the Americas. The popularity of the gloomy thought waxes and wanes with the reciprocal of the price of gold or the positions of Mars and Jupiter or other factors that are difficult to judge with any great precision. And so contemporary Solons of sadness sing their siren songs to students who assume that the message is a sophisticated one. Philosophers, novelists, professors, and self-appointed prophets join together to transfer their burden of angst to the unsuspecting young. The academic pessimists now find their commercial counterparts in the doomsday industry, an over a billion dollars a year market in survival apparatus, supplies, and housing.

I have long felt a profound gut uneasiness about the presence of fashionable despair in the university community. The modishness of gloom seems inappropriate to the enterprise of learning. It has, however, taken me some time to provide a more cerebral setting for these intuitive feelings and to confront the downbeat attitude of teachers and explain why it is so objectionable. Although the problem has been troubling me for many years, a recent reminder of Immanuel Kant's *Critique of Practical Reason* brought the issue back into focus.

It all happened when I was at a cocktail party getting bored and irritated by downbeat small talk. I decided to escape the doldrums by turning to the person next to me and inquiring, "What do you think about Kant's categorical imperative?'' (It's not a bad conversation opener or closer, depending upon whom it is addressed to.) I refrained from asking, deciding it was best first to update my own views on such an ethical dictum. This led eventually to rereading and rethinking the imperative, which has been translated: "Act as if the maxim of our action were to become by our will a universal law of nature." The Kantian statement, which has been influential in most subsequent ethical arguments, would seem clearly to argue against a downbeat attitude. For if these feelings of malaise were to become universal, human interactions would lose that measure of joy that is so precious to most of us.

But beyond the classical argument that has been around for a long time, there is an aspect of the situation that I believe takes on even greater importance. The supplementary idea may be stated as a formal principle: *Because we are participants in the course of human history, the future will not unfold independently of what we believe it will be.* In other words, our thoughts about the future are not intellectual abstractions but active instruments in influencing coming events. Whereas the preceding statements are true for everyone, they are especially true for teachers and writers, whose beliefs are amplified and transmitted to many recipients. Since some of those affected go on to become teachers and authors, the ideas reverberate down through the ages.

Pessimism as a belief not only becomes a passive set of predictions about the future but also plays a dynamic role in ensuring the deteriorating quality of tomorrow's world. If the young are persuaded by intellectual arguments that there is little or no hope, then they are robbed of the emotional energy to carry out the very hard and often frustrating work of building a better society. Thus pessimism is under a double indictment: It fails the Kantian test and then goes far beyond that fault by being an agent of its own fulfillment. Given these considerations, optimism clearly emerges as a moral imperative, an attitude toward the future that is demanded of us by ethical considerations.

The idea that predictions about human behavior influence that behavior has been around in one form or another for some time. The stock market provides some very direct examples. One of these recently involved a well-respected and widely read analyst who predicted a major drop in stock prices. Word of this forecast set off a selling wave that quite naturally drove prices down. We have no way of knowing, in an experimental sense, how the market would have moved in the absence of the economist's actions. Had he predicted the future? Or had he manipulated the future so as to guarantee the success of his forecast? One is reminded of the story of a medieval seer who foretold the exact date of his own death and then committed suicide to ensure the accuracy of his words.

With respect to knowledge of the future, there is a clear distinction between the social sciences and such natural sciences as macroscopic physics and chemistry. When astronomers calculate the orbit of a planet by using the laws of mechanics, there is no thought that they are exerting any influence on that orbit. The accuracy of the prediction is taken as a measure of the validity of the science. Sociology, economics, and related disciplines enjoy no such disjunction between theory and observation and must therefore operate under an ethical burden in making forecasts.

Having, for moral reasons, come out on the side of optimism, a caveat is in order against the followers of Dr. Pangloss's contention that this is the best of all possible worlds. Optimism opens up the chance for future improvements but does not, in and of itself, guarantee that the change will be for the better. The knowledge that an improved world is possible is the first step in the process. There is always the additional requirement for committed individuals who will put forth the strenuous efforts to build and sustain the institutions to deal with the problems of our rapidly changing technological society. The answers lie in the spirit, vigor, and optimism of those people who are both the catalysts and substrates of social ferment.

I must confess that, infrequently, I have fallen victim to gloom and have transmitted some of that despair to my students. Malaise is a hard thing to hide. The pessimism always came with a sense of guilt, for as a teacher of the young, I somehow felt contractually obligated to present the hopeful side of the human condition. The moral imperative imposes an even greater responsibility on those who have chosen teaching as a profession.

Well, you grumps and grouches and dyspeptics out there—your cover has been blown. There is nothing so intellectually deep about your pessimism. Indeed, it could be regarded as a rationalization or easy way out. Your very gloom about the future may be providing you with a reason for not putting forth all the work necessary to overcome the decay that you envision. Recall the words of Henry David Thoreau: "Men will lie on their backs, talking about the fall of man, and never make an effort to get up."

And students, do not accept the sour words of elders telling you that their world is better than yours. Do not believe in the inevitability of social deterioration. You can make of your time what you will, if you will it strongly enough. You have the knowledge that the labor is moral as well as joyous.

Used Theology

There is a special feeling associated with exploring used-book stores. I have often watched browsers avidly take books from shelves and open them. I sense that many of my fellow bibliophiles share a hope that some day they will see hidden between two large tomes a thin ancient volume on which one can barely see the title, *Secrets of the Universe.* They will cautiously extract the book with the skill of an exodontist, thumb through the pages, and "voilà!" But perhaps it is too much to anticipate: God's design for the universe must be sought in other ways.

My recent travels into the out-of-print took me down Telegraph Avenue in Berkeley. Since my companion was my favorite rhetorician it was natural to seize a volume entitled *The Rhetoric of Science* by William Powell Jones. The subtitle, *A Study of Scientific Ideas and Imagery in Eighteenth-Century English Poetry,* awakened and titillated my two-cultures instinct. Moe's ended up with some money and I was the possessor of a fine used book. The good news from Mr. Jones is a recounting of the love affair between science and poetry that characterized eighteenth-century England.

In the immediate post-Newtonian era, there was a poetic euphoria celebrating the orderliness of the universe and the beneficence of a creator who established such divine order. Alexander Pope extolled the revelations of Newton with the wonderful couplet

> *Nature and Nature's laws lay hid in night:*
> *God said, Let Newton be! and all was light.*

Beyond Newtonian physics the microscope and telescope provided views of the world that inspired the authors of verse.

All of science was fit subject matter for the eighteenth-century poets and a point of view emerged that has been termed physicotheology, the use of scientific discoveries to show the wisdom of God in nature. The underlying theme was the argument from design, stated in rhyme by Moses Browne.

> *If lesser proofs such demonstrations show,*
> *What may th' unmeasured universe bestow?*
> *Think, cou'd blind chance, dead unexisting name,*
> *Produce such order, so complete a frame?*

This line of reasoning was adopted by prose writers as well and played an important part in the literature and theology of eighteenth-century England. In France, matters had taken another turn. Followers of René Descartes, looking at the orderliness of nature, tended to minimize God's role in a system that worked so well by itself. That movement led to the publication of the famous *Encyclopédie* (1751–1777) under the direction of Denis Diderot. The view of the world that resulted in the physicotheology of the English led in France to writings considered sufficiently tainted with atheism to be "subjected to Jesuit censorship and royal repression."

At this point in my consideration of the argument from design, I ran out of material from used-book stores and had to retreat to a musty library corner for the quintessential scientific example of that theological line of reasoning. And sure enough, I located eight volumes of the Bridgewater Treatises, published in the mid-1830s. Some of them were last stamped with a withdrawal date thirty years ago. I was not in the mainstream of modern thought.

The Right Honourable Earl of Bridgewater by his last Will and Testament directed Eight thousand pounds . . . that persons should be appointed to write print and publish one thousand copies of a work On the Power, Wisdom, and Goodness of God, as manifested in the Creation; illustrating such work by all reasonable arguments, as for instance the variety and formation of God's creatures in the animal, vegetable, and mineral kingdoms; the effect of digestion, and thereby of conversion; the construction of the hand of man, and an infinite variety of other arguments; as also by discoveries ancient and modern, in arts, sciences, and the whole extent of literature.

The eight authors were individuals of great distinction; five were members of the Royal Society. Each produced a definitive work of contemporary science pointing to the design and order behind all phenomena. While they were penning their words, Charles Darwin was a-sea. He and many others were developing the scientific information that by the 1860s would lead to violent confrontation with the followers of the Reverend Francis Henry, the Earl of Bridgewater. By the end of the nineteenth century the argument from design, poetically hailed 150 years before, had faded into oblivion. In addition to not foreseeing Darwinism, the post-Newtonians had, curiously, neglected an important idea from the past: The God of the argument from design is not the God of the Old or New Testament.

In 1670, Benedict Spinoza had written "Of Miracles," which was part of his *Theologico-Political Treatise*. He stated

That nature cannot be contravened, but she preserves a fixed and immutable order. . . . Therefore miracles in the sense of events contrary to the laws of nature so far from demonstrating to us the existence of God, would on the contrary lead us to doubt it.

This line of reasoning was overlooked by the physicotheologians, who ended up with a kind of pantheism: universal intelligence or world mind. The God of Judeo-Christianity interacts directly in human affairs and therefore exists outside of nature. The post-Newtonians, in introducing the argument from design, were ignoring the profoundly historical epistemological foundations of Western religion and were thus undermining the basis of the theology of the Church of England, which they wished to support. Identifying God with nature can lead to the piety and belief of Spinoza but cannot lead to the God of history, who interacts directly in the affairs of humans with little concern over the fine points of obeying the laws of physics.

Thus, the post-Darwinian agnosticism of Thomas Huxley had its foundations in the religious writings of the eighteenth-century poets and early nineteenth-century divines. The Earl of Bridgewater's bequest ironically served in the end to weaken traditional Western religion rather than to strengthen it.

I recall an experience I had a few years ago in the venerable bookstore Blackwell's of Oxford. I was wandering from section to section and paused to look up at the subject signs. There across from me were the jarring words, "Used Theology." I envisioned an almost new volume that had been read only by a small elderly person on Sundays. Alternatively, it may be the theology that was secondhand, lifted by some hack scholar from the writings of Thomas Aquinas or Maimonides. I suspect that the Bridgewater Treatises would nowadays end up in Used Theology. That is too bad—for the argument from design is a powerful one.

So once again I'm out on the streets, wandering through the stalls of used-book sellers. I refuse to give up the thought that one day for just a dollar and a quarter or so it will *all* be revealed. In any case, the lesser treasures that are uncovered usually serve to justify the endeavor.

The Sphinx's Secret

O ne of the joys of traveling is the opportunity occasionally to gain new insights into a civilization or culture one has long known only through books or films. Novel thoughts may come suddenly, or they may dawn on one many months afterwards.

My puzzlement about pre-Ptolemaic Egypt started many years ago when I was visiting the British Museum. Why had such an obviously dynamic society invested so much of its productivity in pyramids, sarcophagi, and other death art? Years later, the same question came to mind one evening when I was standing on an upper floor of the Ramses Hilton looking out at the Nile. A few minutes before, I had first viewed the scene around me and mused, "Being an academic isn't all that bad." A day of wandering around and crawling through pyramids strongly impresses one with the magnitude of those tombs and the vast amount of human labor that went into their construction and decoration.

That evening I turned down an invitation to attend a sound-and-light show at the Great Sphinx. Some of us have trouble mixing cultural themes. Recovering the thought of an earlier civilization is difficult enough for me without experiencing the artifacts bathed in the intense electromagnetic and sonic vibrations of our own age.

I spent the next day entirely in the Egyptian Museum in Cairo, and I underwent a total immersion in 3,000 years of culture, from the prehistoric to the Ptolemaic period. New information came so fast that it has taken a long time to digest it, and only now, over a year later, am I coming to grips with an understanding of why one such highly sophisticated culture after another seems to have centered so much of its artistic creativity on death for such a long period of time.

My initial impression, gathered at the British Museum, now seems wrong. Seen in the stark displays in London, all thirty dynasties seemed to be engaged in necrolatry, a trade-off of the joys of this life in favor of a morbid concern with after-death experience. But I came to realize that the tombs—set on the sunlit shores of the Nile and filled with food, drink, games, dancers, and attendants—bear witness to a passionate love of life. The death cult sprang from an unwillingness to face the prospect of ever ending the material existence. Through all of these often very different dynastic cultures is woven a common thread, the desire for an afterlife filled with the best of the sensual pleasures of this life.

The ancient Egyptians were not death worshipers. If they erred, it was in their failure to realize that the activities we cherish are inextricably bound to the fragile body, which will not survive death. The classical Egyptian intellect was overwhelmed by an enormous quantity of wishful thinking.

Wandering through the artifacts of thirty dynasties, one is tempted to try to put things in order, to see a pattern over three millennia. It is difficult. I am charmed by the seated Ra'hotpe and his wife, Nofret; awed by King Kha'fre; impressed by Amenihotep, son of Hapu. But placing it all in chronological order and trying to see a temporal development are beyond me. Too many influences have been at work; too many invasions, conquests, and catastrophes have occurred; my lack of knowledge is vast.

And so I trace back the artifacts in time. The displays are arranged chronologically to make this time trip relatively easy. Just before the early dynastic period, there is a radical change in the character of the artifacts. The great statues, tombs, and sarcophagi are replaced by pottery, stonework, and flint javelin heads. Crude figurines accompany wicker baskets and some copper objects.

The museum has only a few materials from the prehistoric Badarian and predynastic periods. Yet the impression

I get is of a very rapid transition, within perhaps two hundred years or less, from a scattered Stone Age culture to the first dynasty, which ruled a well-organized state with an efficient bureaucracy, a well-developed religion, and a powerful military. Architecture, sculpture, and other technologies were the work of obviously expert artisans, and I am being treated to the remains of their skill. Five o'clock approaches, and I have to leave the museum. In retrospect, I see that one riddle has been replaced by another. I entered wondering why thirty dynasties had been so obsessed by death, and on reflection I have some feeling for the answer; but I leave wondering how a Stone Age people could, within two hundred years and without much apparent outside influence, have reached such a high level of material and artistic culture.

There was no time to seek an answer in Cairo, for my plane left at 10 P.M., and I was off to New York. Now, some time later, I'm at the library to seek more information about the first dynasty and the first king, Menes. The sudden emergence of Egyptian culture is well known, and the *Encyclopaedia Britannica* reports the following:

The first unification, therefore, was felt to be so important that it liberated great creative forces in the land. Overnight, as it were, Egyptian civilization achieved maturity; and in its physical manifestations it revealed the characteristics that were to be constantly visible for the next three millennia.

That is a description, but not an explanation, of what I observed at the museum in Cairo.

Perhaps I am taking a diminished view of two hundred years. The United States, after all, went from a group of colonies to a world superpower in a shorter period. The difference is that in the rise of America, technology was at a high rate of development, innovations were constantly imported from around the world, and the immigrants brought with them the sophisticated cultures of their homelands. Perhaps because we have so many more details, it is easier to envision the growth of a modern state than the rise of dynastic Egypt. It is unlikely that we will ever resolve the considerable mystery that persists about the rise of the first dynasty.

The several transitions from Stone Age culture to great empire that have occurred may all have had a somewhat explosive character. If true, that presumably has something significant to tell us about human society if we can understand the dynamics of such changes. I entered Egypt with one riddle and left with another. At the very every least, that provides a reason for returning, and I am mindful of a bumper sticker I saw on the way to the airport. It read "PEACE-TOURISM."

The Sinai Connection

It was one of those unplanned vacations, which add the dimension of surprise to whatever else one experiences. I was in Jerusalem attending a meeting on my favorite microorganisms. At the travel desk, Lynn responded to my inquiry by telling me she could arrange two days of trekking in the Sinai desert. I didn't have time to go into details, and early the next morning I was on an Arkia flight setting down in Eilat, the southernmost city in Israel.

My somewhat cryptic instructions were to go at 9:30 A.M. to a travel agency at the center of town and meet the representative from Neot Hakikar. When I arrived, I got the disappointing message that the trip would not be leaving until 11:30. I deposited my luggage and took advantage of an unexpected two hours to tour Eilat. Eilat was not crowded with tourists. Perhaps the fact that it was almost July accounted for the ease with which the travel desk could get me booked into the desert on such short notice. Never mind. I saw downtown Eilat, purchased new sunglasses, and returned to meet the representative of the trekkers.

We were shortly off to Taba, on the Israeli-Egyptian border a few kilometers south of Eilat. At the Israeli station we filled out forms, had our passports stamped, and proceeded to the checkpoint. The passage from Israel to Egypt had two aspects: the full seriousness of a bureaucracy at work and the profound realization that things can really change for the better, even in the Middle East.

The last checkpoint in Israel was staffed by two young policewomen. Armed only with a telephone, they sat in the shade of a small booth and waved to us as we drove up to the Taba gate, a two-door, wire-mesh, six-foot-high national border, one door of which was open. Three unarmed Egyptian soldiers sat by the fence as we entered on foot, carrying our luggage. The guards waved us on to a row of long, narrow one-story-high buildings. At the second one we were given forms, which we dutifully filled out, and then one at a time we entered the office of a stern official who carefully checked every answer, initialed the form, stamped our passports, and sent us on to the next building, which turned out to be a bank. Here we converted $6.00 into 480 Egyptian piasters, a procedure that required an elaborate form, one copy of which we were given. Then it was on to the tax station, where we returned our bank document and paid 410 piasters entrance tax. I still have the remaining 70 piasters, a souvenir of the ongoing incomprehensibility of the bureaucratic mind.

Out of that building, and still another awaited us. Customs consisted of a man who reached a hand into each suitcase. My luggage, one small shoulder bag, engendered some comment about a book on top of the other contents, but I'm afraid the meaning got lost in the translation. The volume was a paperback edition of *Wanderings*, by Chaim Potok, which seemed appropriate enough reading for trekking the Sinai. You may think that by this time we had fully entered the Land of the Pharaohs, but yet another official made a list of our names and one last functionary rechecked our passports before we moved on to see our "home" for the next two days, a four-wheel-drive Toyota with two bench seats for the passengers.

Far more important than all the paperwork was the cordial and sometimes friendly relationship between the Egyptians and the Israelis; thank you, Anwar Sadat. The trek is apparently a binational joint venture; the truck was Egyptian, as was the driver, a Bedouin named Ahmed. Hassan, an Egyptian guide, worked along with two Israelis, Daniella and Charley. Ahmed's ancestors had ridden camels across this desert long before there were roads, and there were times when he navigated his Toyota in the same spirit. The only difference was that the truck had a horn, which the driver dutifully honked at goats, camels, and any Bedouin who came within sight. Sighting distance of the black-clothed Bedouin women in the desert is about two miles.

We all climbed in and were off down the route along the Gulf of Aqaba. A short distance under way, before we lost sight of the hotel at Taba Beach, it occurred to me that we had within view Israel, Egypt, Jordan, and Saudi Arabia.

Had I the spirit and literary skill of Chaucer, I could describe my traveling companions on the road to Mount Sinai. But, alas, I lack that bard's interest and insight in exploring the life stories of the pilgrims. There were eight of us, including two very quiet, nice Finnish ladies, a pleasant Swiss lawyer working for the Red Cross in Jerusalem, an American couple, and a German couple. The men of the two couples were archetypal examples of the ugly American and the ugly German. They both talked incessantly. They later discovered they had a language in common, Spanish, but they did not seem to enjoy talking to each other because one was then forced to be a listener part of the time, a difficult role for each of them.

I shall describe these two thorns briefly to get over the unpleasantries, because the entire experience was so wonderful it was worth putting up with even David and Manfred. The former was about thirty years old and could talk about any subject as long as it had no content. He kept introducing out-of-place and sometimes incorrect Yiddish phrases into his conversation in what I suppose was an attempt to identify with Israel. His efforts failed for the Jews, Christians, and Moslems among us, so at least his lack of taste had something cosmopolitan and universal about it. I last saw this charmer a week after our trek, when I spied him at John F. Kennedy International Airport pushing a frail white-haired woman out of the way to get his luggage. Manfred, our other gem, was a tall, dark-haired, athletic man, probably in his mid-thirties. When not talking compulsively, he was compulsively taking photographs. To explain this man, I must go a little out of sequence to the time when we were climbing Mount Sinai in the predawn darkness. It was rocky and tricky underfoot. Manfred would run ahead, turn, and take flash pictures, so that we all lost our dark adaptation. At that point I could not resist a terrible thought: "May Yahweh fog all his film, grind sand into his quick-change mechanism, and cause his telephoto lens to fall between a rock and a hard place." But enough unpleasantries.

The border crossing was sufficiently time-consuming that it was soon lunch hour, and we veered off the road to a Bedouin house about 50 meters from the shore of the Gulf of Aqaba. Here we enjoyed a typical Middle Eastern meal of olives, tomatoes, cucumbers, hummus, cheese, tinned fish, and bread. Then after a short rest we went for a swim in the clear, warm, refreshing waters of the gulf. A swim never felt better nor have I ever seen clearer water. The dressing areas were the spaces behind any rock large enough to shield one from the other travelers' line of vision.

The Sinai is not a desert in the sense of unending expanses of sand but rather a wilderness of rocks, mountains, and crags characterized by an almost perpetual dryness. A series of geologic upheavals along the great rift line presents limestone, sandstone, and coral, along with sand produced by erosion of the rocks. Occasional desert shrubs and grasses come up where small amounts of water collect, but very little green is in view; brown and red and gray are the predominant colors of the Sinai. The land is hard on the feet but not unpleasing to the eye.

After the swim, we were off on a long and often very bumpy ride to a rather substantial village at the foot of Mount Sinai. We arrived just before sunset of the last day of Ramadan, so festivity was in the air. We occupied a large flat piece of land and were all given sleeping bags. A tank of propane with accessory apparatus was brought out, and

Turkish coffee and tea were soon available. Hassan began to make a rice dish, Charley began to cook up a stew of some unknown frozen meat (probably veal or goat, but I didn't ask), and Manfred's wife and Daniella started the salad.

After two hours spent watching the mountain gradually change color and then disappear into the darkness, we sat down to eat a most enjoyable meal. A few bottles of wine mysteriously appeared, and those among us for whom it was permitted indulged. We then went off to our sleeping bags, and I lay back looking up at a star-filled sky whose grandeur was too transcendental to be made verbal. In the excitement I thought sleep would be difficult, but the next thing I remembered was the sound of the propane burner heating up the morning coffee and tea. In the darkness, I crawled out of the sleeping bag, rolled it up, and got my coffee. While sipping the strong sweet drink, I heard the Moslem morning prayers begin over the neighboring village's loudspeaker.

We left everything behind except what was needed for the climb and piled into the truck. Ahmed, may Allah cause his Toyota never to stumble, drove us unerringly to the beginning of the trail. It was a little after 4 A.M. when, aided by flashlights, we began our climb. It was a hard walk, and it included a 600-meter vertical ascent along a rocky trail of 700 steps that had been placed by the monks centuries ago to aid pilgrims making the climb.

After about thirty minutes it was light enough to dispense with the torches, except when we lost our dark adaptation. In another half hour it was light enough to pick up the pace, and the group split into two parties of varying speeds. By selecting an intermediate rate between the two I was able to climb alone, thus avoiding both Manfred's guttural staccato fusillade and David's whining vulgarities. In truth, I very much wanted to be alone. I did not expect to see a burning bush along the way nor did I anticipate a voice out of nowhere; nevertheless, I wanted to be in the right state of mind should by chance anything unusual occur. And as an old reader of William James (*The Varieties of Religious Experience*), I was sure of one thing: Solitude is a precondition to that mental state.

I don't have any idea how long the climb took; indeed, I had an acute loss of time sense for that entire day, but eventually I reached the top. I wanted to be alone and I still cannot quite share my thoughts on the ascent. Suffice it to say that for those who choose to imagine, climbing Mount Sinai is going to be pretty heady, regardless of one's metaphysical orientation.

After sitting at the peak and looking at the surroundings now brilliantly lit by the mid-morning sun, we descended about halfway down the mountain to a grove of cypress trees in the middle of an otherwise dry vale. It is presumed to be where the prophet Elisha spent his final days, and it was certainly the home of St. Stephanos the hermit, as well as a number of others living in religious isolation over the ages. Here we had a breakfast of the ubiquitous bread, cheese, tomatoes, cucumbers, and vegetable spreads, washed down with water from the Well of Moses. We each found a relatively flat rock and a little shade and lay down to rest for an hour or so. The sun was bright and hot. At this point everyone was quiet, and the solitude was appreciated.

For the trip down we again split into two groups. David's wife—may Yahweh grant her the patience of a seventeen-year locust—had become ill on the ascent and took an easier descending trail, accompanied by her husband and Charley, who turned out to have been a Belgian doctor in his previous life. The rest of us followed a rather steep, rocky route that led directly to the monastery of Santa Katerina. The monk at the entry complained that the women were not sufficiently well-covered, even though we had stopped to put on long pants and shirts at the base of the trail; the sleeves were not long enough. We toured the monastery, which I found interesting enough but anticlimactic after the top of Sinai. Of course, almost anything was going to be a letdown after climbing to the pinnacle of *that* mountain, and I was lost in reverie for the rest of the day.

For those who are interested in monasteries, Santa Katerina is run by a Greek Orthodox order and has been in continuous operation for over fourteen centuries. It contains exceptional iconography and a very important collection of ancient documents.

The second group of hikers had not yet made it down when we finished touring the monastery, and we were becoming a bit concerned. After some miscommunication we went back to the campsite and then off to a second rendezvous point at a desert "motel." We found that someone else had gone to the monastery to check things out; Ahmed announced that "Allah will take care of everything." Fifteen minutes later, a truck pulled up with the missing hikers.

What with getting up at 3:30 A.M., the tiring climb, the heat, the bright desert sun, and the experience, I felt pretty spacey all the way back. We drove from the monastery to Taba, stopping for lunch and a visit to a Bedouin village. The return trip is but a bumpy blur of a memory.

Back in Eilat it was necessary to face the real world, and I found a room at the Adit, a small, pleasant two-star hotel a block away from the beach. I made my transition from desert dweller to townsman in three steps: vodka and orange juice over many, many ice cubes, then a long, long soak in the bathtub, and, finally, clean clothes. It was late in the evening when I emerged for a slow stroll through the new tourist center. At a food stand I sat down to a mixed-grill sandwich and a bottle of wine. There were few customers around and I invited the proprietor to join me for a drink. We sipped wine and talked of this and that for an hour or more. I wandered back to the hotel and settled down to the best night's sleep I had had in a long, long time.

Who, When, Where?

I n one of his perceptive essays on culture, Erwin Schrödinger expressed doubts about any scientific enterprise that wasn't directed at the questions: Who are we? Where did we come from? Where are we going? The second of these questions has undergone a rather extensive change of character since the development of molecular biology, and the answers being offered are as diverse as, if less acrimonious than, they were in the days of the Huxley-Wilberforce debate on evolution. The problem of our origins is thus intimately related to our identity since we are embedded in an evolutionary history that ties these two issues very closely together.

The present philosophical viewpoints on the origin of life can be classified in the following way:

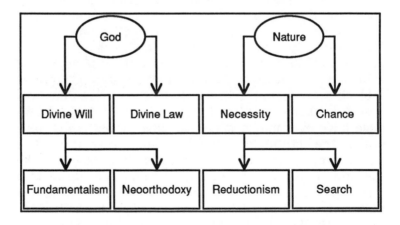

The theologically centered views themselves are very broad. They divide into the divine-will position of traditional Western religion and the divine-law view that began with Spinoza and can be seen in the works of recent writers, such as Pierre Teilhard de Chardin. The divine-law school overlaps strongly with the necessity school of scientific thought, and one feels that they may be separated more by style, tradition, and syntax than by basic philosophical issues. The divine-will group includes holders of two outlooks: fundamentalists, who believe that all is known about the origins of life and is written in scripture, and neoorthodox existentialists, who believe such matters to be ineffable mysteries.

The particular division of nature-centered views outlined above stems from Jacques Monod's controversial book entitled *Chance and Necessity*. Monod states that our origins cannot be traced by the laws of physics because so many very low-probability events have occurred along the way as to make a scientific analysis impossible. While he believes that the phenomena of life are always reducible to physics, he asserts that they are not predictable from such considerations because of the vast array of highly unlikely events that must have led to us in our present state. Thus to Monod, the molecular biologist, life's origins are just as much a mystery as they are to the neoorthodox. One uses the word "chance" and the other uses the words "divine will," but they both describe a series of events that, by their uniqueness and unknowability, lie outside science.

The hard-nosed school of chance-oriented molecular biologists and the neoorthodox would find themselves strange bedfellows, yet they are tied together by their existentialist roots. Each looks across a gulf of unknowable mystery. The theologically oriented sees God on the other side of the gulf; the secular existentialist sees a vigorous chaos churning up such unlikely events as you and me.

As a scientist, I take strong objection to the Monod point

of view. To assert that something is unknowable is to discourage those experiments and theories that would lead to a change of paradigm, converting the unknowable into something more tractable. Self-fulfilling prophecies have no place in science since they impede the free development of ideas. Negative statements, such as the second law of thermodynamics and the Heisenberg uncertainty principle, are, of course, part of the structure of modern physics, but they arise out of experiment and construct development rather than standing as barriers to these activities. Chance has at least two roots: one, a basic characteristic of classes of events; the other, our lack of knowledge of sufficient details. Since it is often difficult, if not impossible, to distinguish between the two, it is intellectually dangerous casually to commit whole classes of events to such a know-nothing category.

Advocates of the necessity view divide into those who think that our origins are understandable within the structure of contemporary physics and those who search for further scientific principles to understand the self-organizing character of matter under the influence of energy flow. Both schools of thought see the origin of life as flowing in a spontaneous way from the unfolding of the universe. They assert that neither were we willed into being nor did we randomly stumble into being; rather, our being is a natural part of the evolving cosmic processes—we are at home in our universe. This view does not remove all mysteries. The existence of existence still stands as an impenetrable barrier. However, we need not be so presumptuous as to assume that all veils will be removed to reveal the naked truth before us. In return for our incomplete knowledge, the joy of exploration still awaits us. Such a search has rewards of its own.

Among the scientists who believe that the present apparatus of physics and chemistry is adequate, there are two distinct groups of research activities: The first school stems from organic chemistry and biochemistry and goes back to the experiments of Harold Urey and Stanley Miller, which demonstrated the synthesis of amino acids and other compounds of biological interest in an atmosphere of ammonia, methane, and water, subjected to a spark discharge. This type of random synthesis experiment has now been repeated under different conditions thousands of times, yielding an inductive generalization that high energy input into a reducing atmosphere of compounds of carbon, hydrogen, nitrogen, and oxygen will produce a class of compounds that includes many biochemical intermediates. The second school has its origins in irreversible thermodynamics and asserts that the flow of energy leads to dissipative structures that order a system and such ordering will lead to biogenesis. Crucial to this viewpoint are the ordering tendencies in far-from-equilibrium systems.

Both reductionist views have a tendency to present scenarios of life's beginnings that go far beyond what can be said with certainty and, surprisingly, seem to interact little with each other, leading to two almost noncongruent scientific views of the subject. The split view leads me to believe that neither school has really got to the heart of the matter.

Life as we know it is a series of cyclical processes involving macromolecules of exquisite precision and great structural specificity. Life also is tied to an unending processing of energy to continually repair the degradative ravages of thermal decay, the ceaseless tendency toward entropic increase. The random synthesizers approach the problem from below, working up from the properties of atoms; the thermodynamacists approach the problem from above, using analysis of macroscopic systems. The two do not meet. Neither extends far enough into that intermediate hierarchical level that is the core of the living process. When we move into that domain, both classes of theories seem too structureless; they lack the ability to deal with the steric factors and the networks of chemical flows. Existing models appear too crude, too coarse for the eloquence of macromolecular structure and function that characterize life.

Worded in a more technical way, quantum mechanics appears conceptually inadequate to problems of biological complexity, and thermodynamics is incapable of generating structures—it is only able to operate with structures that are presented phenomenologically.

Given the inadequacies of contemporary approaches and the existence of life challenging us to a scientific explanation of its origins, we conjecture the need for at least one additional principle to bridge the gulf. This is treading on dangerous ground because twentieth-century scientists have been very unkind to thinkers who have proposed the introduction of biogenic laws. The tendency has been to excoriate them, after first labeling them with the epithet "vitalist."

Henri Bergson was perhaps the first in this century to propose that biology demanded laws that went beyond the physics of his day. More recently, Walter Elsasser has pointed out with great thoroughness that the laws of quantum mechanics, dealing with classes of identical entities, are inadequate to a biology that deals with entities whose complexity renders the concept of identity meaningless. Elsasser has been subject to considerable criticism by molecular biologists who have not understood the depth of his message.

It is rather interesting from a historical point of view that attempts to introduce new physical principles to deal with biology are met with such opposition. We regularly extend physics into the high-energy or low-temperature domain with the introduction of new appropriate relations. Yet the call for a new physics to deal with the range of complexity appropriate to biology is met with opposition. Why do we always assume that physics may be incomplete in all other spheres but is complete with respect to biology? I suspect that there are two main reasons for these attitudes. First, modern biology has been so eager to exorcise the demon of vitalism that there is an uncritical attitude about just what constitutes vitalism. Second, many biologists have been sufficiently unaware of the nature of physics as to assume that it is much more complete than it really is.

In any case, if we are going to recover our origins, we will have to go beyond present theories. A possible search for biogenetic laws is before us. The theory need not lie outside physics; it can extend physics into that domain whence we can better face the question of where we have come from. It is important to realize that science is a very young enterprise. We are probably only at the beginning of our understanding of the world around us. It is somewhat arrogant to assume that all we do not know is unknowable; rather, we should bend our efforts toward increased understanding. It seems clear that further search is the only path to deepening our insight.

I must confess that it is sometimes uncomfortable to stand between the reductionists and fundamentalists, who believe that all is known, and the religious and secular existentialists, who believe that all is unknowable. All these proponents seem so sure of themselves that those of us who believe in the search as a way of life seem equivocal in comparison. If the search, with its uncertainty and insecurity, is emotionally less satisfying, it is intellectually more open and can lead from the present position to an ever-deepening understanding of the questions I raised at the outset. For some of us, no other path seems possible. Indeed, amid the present search for "roots," seeking out our molecular ancestry seems like a very appropriate activity for those who believe that such beginnings are knowable.

Reality, Reality

I am about to lose my wavering liberal credentials and to alienate some individuals whom I would much rather have as friends. The search for truth imposes harsh responsibilities and now impels me toward a task I do not relish. It all began in May 1983 at an American Association for the Advancement of Science meeting in Detroit. While browsing through the book displays—my favorite activity at meetings—I happened upon *Discovering Reality: Feminist Perspectives on Epistemology, Metaphysics, Methodology, and Philosophy of Science,* edited by Sandra Harding and Merrill P. Hintikka (D. Reidel Publishing Co., Hingham, Mass., 1983). My on-site reading left me puzzled; now, several months later, I am exploring the book at leisure, and I am disturbed at what I find.

Before commenting on the book, I must tell the story of a debate between a philosopher and a psychoanalyst. The philosopher questioned the epistemological basis of Freudian theory. The analyst looked at him with a puzzled expression and said, "Why do you hate your mother?" Similarly, this book has a built-in defense against any criticism, for such comments can come only out of the "sex gender system," which the authors seek to eliminate. Within their framework there are no arguments opposed to their views that are not structured by "the patriarchical unconscious" and thus are "phallocentric." To predemolish one's opponents with such epithets may be great fun, but it constitutes a kind of know-nothing approach to intellectual discourse.

A gender-free description of this book shows that it is 332 + xix pages, consists of 16 articles plus an introduction, and is multiauthored. Any further statements will not be gender-free.

Two usually tacit assumptions underlie most of the articles:

1. Reality is more private than public.
2. Sexual dimorphism at the cerebral or intellectual level is so great that reality at the epistemological and ontological levels will be gender-dependent.

The first assumption is legitimately arguable and has been a central issue in philosophy since Locke, if not earlier. The second is so blatantly sexist and so lacking in supporting evidence that one wonders how it can be set in print as a serious point of view. The very core of reality-centered philosophy is to define and discuss those issues where our common humanity overcomes our individual differences, whatever they may be.

This book avoids the questions relevant to our collective existence by not dealing with the subject matter that has commonly been designated as epistemology, metaphysics, and philosophy of science. Let us take epistemology as an example. The current philosophical approaches began with René Descartes, involved David Hume and Immanuel Kant, and continue in our century in the work of such philosophers as Karl Popper, Susanne Langer, Henry Margenau, and Bertrand Russell. The general issues addressed by these authors hardly ever appear in this book. Rather, the first two chapters, for example, inform us that Aristotle was a sexist and women are severely economically disadvantaged. The conclusions are true and interesting and the situation deplorable, but all of this is quite unrelated to the title—as is the article "Charlotte Perkins Gilman, Forerunner of a Feminist Social Science."

Equally removed from the title are two chapters pointing to the sexism in evolutionary thought. Article after article, the book makes it painfully clear that women have been disadvantaged and viewed unfairly by the male intellectual community. Who can doubt it? But on the other hand, need we suggest that the Pythagorean theorem, Newton's laws, the second law of thermodynamics, and Schrödinger's equation are gender-dependent? We have achieved much in science because of an egalitarianism of ideas. To suggest two kinds of basic science is to attack the equality that has made the discipline possible. Are we now being asked to believe that Pierre Curie and Frédéric Joliot-Curie had gender-related views of physics different from those of Marie Sklodowska Curie and Iréne Joliot-Curie?

If we were to accept the postulate that the epistemology of science is gender-dependent, wouldn't we then have to assume that it is race-dependent, religion-dependent, age-dependent? Wouldn't the whole of science then lose its public character and dissolve into a myriad of solipsistic endeavors?

If one recognizes the absolute moral imperative of equal rights, equal wages, and equal opportunities for women, is it necessary to go that further step to reclassify men and women as *unequal* in their most abstract and most philosophical activities? Having struggled to rid ourselves of sexism in the marketplace, need we reintroduce it at the lab bench and seminar table? A distinction needs to be drawn between the sexism of male scientists from Aristotle to the present and the objectivity of the disciplines. If early Darwinians viewed evolution wrongly because of their Victorian bias as to the relation of the sexes, they were doing bad science. The faults of the practitioners should not be attributed to the intellectual framework, particularly at the level of its foundations.

The extremism implied in the subtitle of this book is an example of a good idea gone mad. Most of the articles are more temperate analyses of how male scientists and philosophers of the past have been outrageously wrong in their views of women. Many of these essays are well documented. Had they appeared under the title *A Sociological Analysis of the View of Women in Science and Philosophy,* they could be recognized for their intrinsic value. The ornate heading under which these writings appear is deceptive.

The overly intellectual title leads to some really silly sentences, such as the following: "The apparently irresolvable dualisms of subject-object, mind-body, inner-outer, reason-sense reflect real but repressed dilemmas originating in masculine infantile experience of the sexual division of labor."

I'm exhausted from reading this book. Statements like that one have caused me sleepless nights. I find it difficult to live with the knowledge that I am "phallocentric" with a " patriarchical unconscious," although I readily admit to being caught up in the "adversary paradigm." Why is that gender-related? Some of my best adversaries are women.

The Bottom Line

For those of us outside the discipline of high-energy physics but still interested in fundamental questions, the past few decades have been puzzling and often frustrating. We are told that our colleagues are closing in on the underlying structure of the universe, and there is baffling talk of "strange quarks," "color forces," and "intermediate vector bosons." One reads that the secrets of cosmology and the history of the universe lie within a unified theory. We wonder if there is a bottom line to this "most fundamental" branch of natural philosophy. These words are used with some trepidation, for there is within physics a tradition, going back to Erwin Schrödinger, if not to Bishop Berkeley, that mental is fundamental. But even putting this argument aside, it is hard to ignore the epistemological issues raised by contemporary physical science.

While pondering these problems, I noted with pleasure the announcement of a forthcoming popular lecture by Emilio Segrè, discoverer of the antiproton. I had just finished reading his book, *From X-Rays to Quarks*. Segrè, whose career coincided with the growth of modern physics, had written, "Scientific research is just as fascinating, dramatic and full of human interest as artistic creation." I decided to attend the lecture.

Segrè began by considering events of the late 1800s, the discovery of the electron and the proof of the atomicity of matter. The apparatus of that era was relatively simple, much of it designed and built in the laboratory. Banks of batteries to provide electricity were the most expensive and demanding equipment. The experimental device used by Ernest Rutherford to observe induced nuclear disintegration consisted of little more than a gas chamber, a radioactive source, a fluorescent screen, and a microscope.

At the beginning of the 1930s, there began in nuclear physics a program, which continues as the main conceptual and experimental approach. A homogeneous beam of particles (protons, alpha particles, atomic nuclei, electrons, etc.) is accelerated to a very high speed and impinges on a target, which is usually composed of atomic nuclei. In a variant of the scheme, two beams of particles are accelerated toward each other. The bombarding particles collide with the targets, producing a new set of particles, resonances, and events that are detected by photographic films, bubble chambers, spark detectors, and other devices designed to record the results of the interactions. The first of these sophisticated detectors was the cloud chamber of C.T.R. Wilson, who apparently conceived the idea from his interest in Scottish mist over fields of heather. The bombardment technique just outlined has been described by physicist Sheldon L. Glashow in the following words:

> An expensive way to find out what watches are made of is to bang two of them against each other and examine what falls out. This is one of the few techniques we have to study subatomic particles and it is for this reason that atom smashers, or more properly, particle accelerators, are built.

For twentieth-century physics there have been three main sources of high-energy particles: radioactive materials, cosmic rays, and accelerator beams. Cosmic rays come from outside our planet and provide a constant source of high-energy events for investigators. These occurrences are relative rare and somewhat unpredictable, thus providing some problems for routine experiments. Nevertheless, much of the early work in high-energy physics was done with this naturally occurring source because it provided higher-energy events than physicists were able to produce in the laboratory. Present-day experiments are dominated by particle accelerators.

The job of the theoretical physicist in the current scheme

of things is to construct theories about the nature of matter that will account for the results of the bombardment experiments. By 1933 this had led to the remarkably simple view that there were three fundamental particles: electrons, protons, and neutrons, the latter two forming nuclei that are surrounded by orbiting electrons. The increase in the complexity of the physicist's picture since 1933 has been vast, exciting to the participants, and hard on outsiders trying to grasp the nature of matter—"the thing in itself" of Immanuel Kant. And indeed, theoreticians are vigorously involved in forming unifying hypotheses to account for the plethora of observations.

Toward the end of his lecture Professor Segrè began to speculate on where all this activity was leading. He noted that in recent years each new advance toward the fundamental was made by building higher-energy accelerators. Higher-energy events keep revealing new and illuminating aspects of the nature of things. Segrè also pointed out that for engineering reasons higher energy inevitably means higher cost. This was of minor importance in the days of "string, sealing wax, and love," when machines were built with much sweat and little money, but today's major accelerators, such as the Fermi Laboratory in Illinois, cost billions of dollars, a nontrivial amount in terms of the gross national product or any other measure of potentially available funding.

Implicit in Segrè's remarks, although he did not actually say it, is the conclusion that in modern physics our notion of "fundamental" is limited by the number of dollars, marks, rubles, pounds, francs, or yen we are willing to shell out. Deep within the structure of contemporary natural philosophy appears the crude capitalistic notion "you get what you pay for," not only in the marketplace, but in the search for the nature of things. "Money is knowledge" seems appropriate enough in the boardroom, but it's a bit jarring within these ivy-covered walls, where the writings of Plato, Hume, and Wittgenstein still make a difference.

I walked across campus, a bit disconcerted about having the ultimate snatched out of our hands because it is ultimately too expensive. I was desperately searching for a way of avoiding a conclusion that seemed so strong. Back in my office, down from the shelf came a copy of Henry Margenau's *The Nature of Physical Reality*, my guidebook in moments of epistemological anguish. This analysis of the philosophical foundations of physics through 1950 listed connectivity, the interconnection of constructs, as one of the essential metaphysical criteria in establishing "physical reality." Our confidence in a theory is enhanced by the number of connections it makes to other constructs. Connectivity has literally been the glue that holds all of physics together as a unified science. The obtaining of Avogadro's number independently by half a dozen different methods is an example of the power and use of this criterion.

Yet modern high-energy physics has found it difficult to establish the degree of connectedness found in the pre-1950 period. Experiment has gone down a narrow road of particle acceleration experiments. Theory is thus in a domain that is difficult to connect with other areas. It is widely assumed that high energy is the path to the fundamental, but like all other assumptions, this too must be questioned. Bombardment experiments may or may not be the only way. Because of the connectivity problem, the philosophical criteria in this area are weaker than in other branches of physical science.

Therefore it is probably too early to conclude that money is the sole answer. Some heroic natural philosopher might yet rescue us from the gloomy economics discussed by Segrè. Money is important, but if we seek the fundamental we had better use some of this cash to feed philosophers as well as accelerators.

Fauna

Overkill

Come, thou mortal wretch,
With thy sharp teeth this knot intrinsicate
Of life at once untie: poor venomous fool,
Be angry, and dispatch.
—Shakespeare, *Antony and Cleopatra*

Among reptiles, the Serpentes are members of a suborder that have been feared, loathed, and worshiped by humans over recorded time and undoubtedly back into prehistory. From an evolutionary perspective, snakes are a relatively young group that evolved from the lizards a scant 80 million years ago. From the point of view of most people I know, snakes have been around long enough, longer by far than *Homo sapiens.*

Among the Serpentes, the particularly venomous family of Elapidae includes cobras, mambas, death adders, tiger snakes, and the less well known but no less deadly kraits. These creatures tend to be indigenous to Southeast Asia and Australia and to find their ecological niches as predators of other snakes. Among these carnivores is found the nocturnal, generally nonaggressive Taiwanese banded krait (*Bungarus multicinctus*). When members of this species bite, they inject a venom at least ten times more deadly per unit weight than cobra venom. Although they rarely bite humans, one recalls the wise words of Ogden Nash:

> *This creature fills its mouth with venum*
> *And walks upon its duodenum.*
> *He who attempts to tease the cobra*
> *Is soon a sadder he, and sobra.*

Our fear of such highly poisonous creatures persists, but

modern experimental techniques now enable us to focus on the biochemical strategies they use in killing prey and to speculate on the evolutionary origins of that interspecies warfare.

All of the prey species, the food of the elapid snakes, are animals that depend for their functioning on muscles under the control of neural networks. The nerves interface with the contractile fibers at neuromuscular junctions. The muscles act specifically on command signals received from the nerves. Electrical signals, or action potentials, are transmitted from the central nervous system to terminals at the neuromuscular junctions. The neural membranes stimulated by the action potentials release pulses of acetylcholine molecules in the neighborhood immediately adjacent to the muscle membranes. The acetylcholine molecules bind to muscle cell receptors and thus alter the properties of the membrane and cause the muscle to contract. The acetylcholine molecules are then broken down by enzymes and reassembled on the nerve side of the junction.

The elapid venoms all contain neurotoxins, molecules that interfere in one way or another with the transmission of signals from nerves to muscles. When the interference affects the respiratory system, breathing stops and death by suffocation results. A whole series of neuromuscular toxins has been studied by toxicologists—the best known probably being curare, the arrow poison of South American Indians, and botulinus toxin, the deadly bacterial agent that sometimes occurs in canned foods.

Let us now reconsider the Taiwanese banded krait and a series of experiments carried out by C. C. Chang and C. Y. Lee in Taiwan in the early 1960s (reported in *Archives Internationales de Pharmacodynamie*, 1963). The experimental section of their article begins, "The venom of *Bungarus multicinctus* used in this study was freshly collected in the laboratory and stored in dried state in a vacuum desiccator." I must confess that this description conjures up a number of vivid images, but no further details are given.

The investigators then dissolved the dried Bungarus venom and subjected it to electrophoresis. The major one of the four components, called fraction II or α-bungarotoxin, binds very tightly to acetylcholine receptor sites on the muscle membrane of the victim and prevents the transmission of signals from nerves. That active agent of the snake toxin possesses an action identical to that of curare, but from a molecular point of view, it bears no relation to the plant alkaloid. Fraction II is a peptide of less than 10,000 molecular weight. Fractions III and IV, called, respectively, β- and λ-bungarotoxin, are closely related proteins that block the release of acetylcholine by a mode of action resembling that of botulinus toxin. Fraction I possesses the action of the enzyme acetylcholinesterase and decomposes existing acetylcholine into its components, choline and acetic acid.

The venom of Bungarus thus operates by a strategy of blocking signal transmission at the neuromuscular junction in three separate and independent ways. It is as if the snake compounded a batch of curare, botulinus toxin, and acetylcholinesterase and injected that mixture into its prey—clearly, a case of massive overkill. One wonders about the evolutionary advantage of such a witches' brew.

When snakes evolved from other carnivorous reptiles, they underwent a loss of legs and developed alternative modes of locomotion, which made available a number of new environmental niches. However, without the grasping forelegs, the problem of capturing prey became more difficult. One group of snakes, the constrictors, developed the ability to coil around victims and crush them to death. A second group, including the elapids, developed toxins of several types. A common feature of many types of toxins is that they are made of chains of amino acids, which are coded directly by DNA base sequences and therefore evolve quite rapidly. The advantage of synthesizing toxins that attack the neuromuscular junction is that such poisons are equally effective on all prey, since they operate on a structure that has, from a biochemical point of view, remained virtually unchanged over hundreds of millions of years of evolution. The weapon changes rapidly, while the target is relatively unchanging.

The alpha neurotoxins are small polypeptides of less than eighty amino acids in length. They are very similar in structure in cobras, kraits, and sea snakes and thus presumably represent a strategy developed early in the evolution of the Serpentes.

Explaining the evolutionary development of venoms with multiple independent toxins seems difficult from a genetic or ecological point of view. Beta toxin by itself is three times as lethal per unit weight as alpha toxin and eighteen times as potent as fraction I. Alpha toxin, however, seems to operate more rapidly. It is not easy to see what there is about this particular mixture that is optimal in the life of the krait. In any case, the complex venom and multiple-toxin strategy used by the banded krait reminds us once again that our view of evolution must be somewhat simplistic. Our paradigms may themselves involve some intellectual overkill. I suspect we have a long way to go to understand the evolutionary significance of the work of Drs. Chang and Lee.

Ego Niches

As a biologist who has spent many years wandering through the complex jungles and perilous reefs of human organizations, I have become aware of surprising and perplexing similarities between the strategies used by individuals to defend their ego roles within hierarchies and those used by animals in nature to defend their ecologic niches. At first sight these correspondences seemed strange and mysterious, but it has gradually become clear that since survival is the name of the game in both cases, a type of convergent evolution is understandable in both the life sciences and the social sciences.

Before taking our field trip through offices and conferences rooms to witness the phenomena in their ecosystems, we must briefly note that many individuals are bifunctional in the roles that they play in their organizations. On the one hand, they are engaged in activities outlined in their job descriptions and seen on the organizational chart; on the other, they are concerned with defending their egos from real and imagined assaults. The ego-defense function is something that individuals bring to work with them from a lifetime of psychological development and may be quite unrelated to their jobs. Indeed, the techniques used to guard the self may be at odds with the functional role, thus causing great difficulty within the company or agency. The growling, snarling, hissing, and predation of the wild can be most disconcerting in the boardroom or at the watercooler. Those ideas can best be illustrated by example, rather than by elaborate theoretical concepts.

The octopus ego niche will provide us with a paradigm of the convergence of organizational and ecological survival strategies. In nature, this normally shy creature lurks along the ocean bottom hiding in rocks, crevices, reef openings, and other protected spots. However, the male octopus in search of food or a mate must sometimes leave his hiding place and expose his soft body and tasty arms to possible attack from predators, such as sharks and eels. Under this threat, the octopus has developed a remarkably clever defense—he squirts ink. The opaque fluid diffuses into the surrounding water and so beclouds the area that the pursuer becomes confused and disoriented, while the frightened cephalopod crawls away to safety, unnoticed in the murky haze.

The institutional octopus, demonstrating the great unity of nature, employs an identical technique. He spends most of his time hiding within the recesses of a small, crowded office, busying himself with his work. When he senses a threat to his tender ego, he too responds by squirting ink. The more haze, the more security—so we see an outpouring of memos, reports, photocopies, computer printouts, graphs, and all the other modern variations of classic ink. The clouds so confound the organization that in the ensuing confusion the octopus retreats into his office to await the next assault. Those who have had to read their way through the output of the human cephalopod will well appreciate the efficacy of ink release as a protective mechanism.

Another marine creature providing a model for an ego niche is the puffer, known to biologists by the tongue-twisting name of Tetraodontidae. Free swimmers in many coastal waters, those animals are often sighted by large, hungry carnivores seeking a meal. In such circumstances the puffer swallows water or air and blows itself up to many times its normal size. The predator decides it does not want to tangle with such a large and important creature and swims off to find smaller fry to fish for.

Analogously, an organizational puffer inflates himself in order to impress any potential challengers so they will go off and leave him alone. This is the niche of the petty bureaucrat who keeps you endlessly waiting while he occupies himself with the "vital" work of shuffling through the papers in front of him. You can sometimes literally see such an individual puff himself up so as to impress you with his full importance. Puffing is also the ploy of the name-dropper, the degree-flaunter, and the inaccessible administrator who is too important to be seen. Who among us has not seen them symbolically swallowing air to impress us with what big men they are? In truth, the puffer is so afraid of his smallness that he devotes himself to developing ego-inflating actions in defense against simple situations. Beware, however, for the puffer in nature is often equipped with poisonous porcupinelike needles, and some of their human counterparts are similarly armed.

An African cobra has evolved the remarkable talent of accurately spitting venom—sometimes hitting the eyes of the victim from as far as twenty feet. The target animal, temporarily blinded and in pain, is totally distracted while the cobra goes about its business without fear of counterattack. The organizational cobra occupies a position of functional importance. Certainly, no one in the institution would tolerate such venomous attacks upon the ego if the victim were not utterly dependent upon the noxious beast for the performance of a necessary task. The occupant of the cobra niche is using an efficient psychological ploy. Everyone else is so busy staggering around, blindly writhing in pain from the cobra's attacks, that no one notices all the faults, errors, and inadequacies. Only the other day I overheard a conversation between two students that contained a perfect description of a cobra: "There's a lady in the Financial Aid Office who is so nasty that you think she's competent."

All animals alter their surroundings somewhat in order to provide a more suitable habitat. The two extreme examples of that type of activity are the human and beaver. The great danger from such species lies in the fact that the alterations that they bring about may make the environment unsuitable for other species. Therein lies the story of the beaver ego niche.

The North American beaver, *Castor canadensis,* is capable of performing prodigious tasks. A husky specimen, using only its big front teeth it can fell a tree five inches in diameter within three minutes—the essence of the beaver's activity is hard work and constant effort. The phrase "eager beaver" has become both a mark of praise and a mild epithet. The reason for the mixed reaction is that the beaver, while working very hard itself, may have managed to dam things up for everyone else. Indeed, we should require beavers to write environmental impact statements before they do a dam thing.

The organizational prototype of these eager animals is, in general, an extremely insecure individual—overanxious to please the bosses in order to preserve his position. As a result, he responds by always biting off more than he can chew, even more than the hundred-foot trees attacked by his natural counterparts.

Oddly enough, the most famous eager beaver of all time did not live in a capitalist country but in the Soviet Union. Alexi G. Stakhanov mined 102 tons of coal in a single work shift. He consistently exceeded production norms, and in 1935 the state began a propaganda campaign urging all workers to imitate Stakhanov. Children joined societies of Young Stakhanovites. It has been reported that the higher production norms led to excessive pressure on other workers and the quality of goods was sacrificed to quantity. For a time the eager beaver niche for everyone was official state policy. Wiser heads finally prevailed, and hardly anyone ever hears of good old Alexi anymore.

Among the largest of the land animals, the rhinoceros is considered the most dangerous by many inhabitants of India and Africa. Possessed of a large horn, erratic temperament, and poor eyesight, this animal frequently engages in ferocious charges in unpredictable directions. There are cases of record of unprovoked attacks on cars, and in at least one case, a train. Since these beasts may attain speeds of thirty miles per hour, they are equivalent to a medium-size truck. Ill-tempered and unpredictable, they may be set off by any unfamiliar sound or smell.

The organizational rhinoceros guards his ego by random and violent attacks. The victim never know what act (a symbolic sound or smell) will provoke the behemoth and so lives in constant fear. The rhinoceros may go for long periods acting like a perfectly calm, well-behaved worker and then suddenly launch forth in an attack on a colleague or employee. The poor eyesight of the natural rhinoceros is reflected by his organizational counterpart's leveling his wrath at any object in the vicinity, without thought of whether or not it represents a real threat. The rush may come in the form of shouting, stomping, desk pounding, and the like, but it always has the characteristics of protection by terror. An organization can seldom tolerate the rhinoceros for too long; we can only hope that this troublesome human species is becoming endangered in modern society. One possibility that seems worrisome is that "assertiveness training" will succeed beyond its proponents' wildest dreams and create a whole new generation of occupants of the rhinoceros niche, who will as a cohort stampede through the ranks of management. We merely note that a word to the wise is sufficient.

To take us to the forefront of research, we may view a niche that has only lately been discovered by scientists engaged in this area of human studies. From the point of view of the work ethic, the cowbird is the most reprehensible of beasts. The innocent-looking creature lays its eggs in the nests of other species, leaving the foster parent to hatch the eggs and rear the young while the lazy parents gad about freely with no sense of responsibility.

Even this bizarre behavior has found its analogue within the organization. The human cowbird moves from office to office depositing his work in other people's in-baskets. He distributes his entire load, then freely spends time as he chooses. Each unsuspecting recipient of such gifts has a bit more work to do, unaware that the cowbird has been entirely freed from responsibility and is therefore un-threatened by the consequences of poor performance. If you think you have one around, keep your eye on your in-basket.

These few examples are chosen from among the hundreds of cases we have assembled in an effort to better understand our colleagues, our employers, and our employees. Without entering into the many disputes over sociobiology, we note that animals have had much to teach us about human behavior since the time of Aesop. Look around you, friends. Is that a frog who jumps from lily pad to lily pad to avoid confronting real problems, or are you looking at a peacock, who builds his or her ego with a gaudy, overdone sexual display? Are you acquainted with a hibernator, who goes into a state of suspended animation when events are unfavorable? Alas, pity the yeast, who fills its environment with alcohol to protect an aching ego, or the baboon, who responds to a threat by grabbing and devouring the nearest food. Have you seen them with a double martini or a box of chocolates? We have all, of course, encountered the organizational giraffe nibbling at the succulent leaves above while his associates stick out their necks for him. Is he mediating about the fine points of economic cycles while the company goes bankrupt? Doubtless, readers could fill an entire zoo with their own examples of animal-like strategists, who are constantly employed in an effort to survive the slings and arrows of outrageous fortune—or at least their own perception of them. Once you get into this thought mode, you can see endless variations of imitations of your animal past, and sometimes even of our far-distant relatives, the plants. There are no limits to human ingenuity in rediscovering protective devices.

After you have grasped the fundamental principle, your relation to your colleagues will take on new dimensions as you categorize the subtleties of the ego niches inhabited by those around you. In the end this will lead to a more sympathetic view of those whose animal-like and plantlike idiosyncrasies are but attempts to guard vulnerable, exposed, and frightened egos. If you are led to more understanding, you will be moving toward the human ego niche, where we recognize and appreciate our humanity, our strengths, and our weaknesses and therefore no longer find it necessary to utilize the devices of tooth and claw to defend our inner selves. If we will make use of those human evolutionary traits of rationality, cooperativeness, and group loyalty, we will come closer to that niche reserved for *Homo sapiens.*

Mitochondria and Matriarchies

In one of his best-known essays, "Organelles as Organisms," Lewis Thomas discusses the theory that mitochondria, the energy-processing centers of cells, are endosymbionts, invaders of a billion or so years ago who have established a mutually beneficial relation with their host cells. He worries about what will happen if his symbionts get a viral disease and how his human dignity is affected by being inhabited by billions and billions of genetic strangers that have their own DNA, totally separate from his genetic material.

Ten years after Thomas's essay first charmed his readers, the study of mitochondria continues to provide us with surprises and insights about ourselves. Because the organelles carry their own genetically active DNA, we can now inquire into their lineage, and it has already been determined that their inheritance is overwhelmingly maternal. These independently replicating entities come almost entirely from the ovum, with the sperm contributing at most a minor fraction of the original population. Maternal mitochondrial inheritance has also been observed in rats, mice, frogs, and fruit flies.

There it is, men! Not only do we have to get along in life with a runty Y chromosome instead of a full second X chromosome, but in addition our contribution to human offspring in terms of informational DNA per cell is significantly less than that of the female of the species.

To reinforce our appreciation of the real importance of mitochondria, recall that there are up to 9,000 of them in each mammalian cell. They are the principal pieces of hardware for converting food energy into its biologically most useful form, adenosine triphosphate. Over 90 percent of all utilized human energy is processed at inner mitochondrial membranes.

When Lewis Thomas learned that a substantial portion of his body was more bacterial invader than human in form and descent, he mused, "There is the whole question of my identity, and, more than that, my human dignity." Now it is more than dignity on the line; a generation of men made to feel guilty by Betty Friedan, Germaine Greer, and Gloria Steinem are now becoming less important as the result of a science that they helped to generate. The assaults on machismo continue unabated.

And do not underestimate the significance of mitochondria; their functioning is crucial in human health. H. S. Rosing and colleagues report in *Annals of Neurology* (in press) that "a large family with familial myoclonic epilepsy was found to have a mitochondrial myopathy which followed a maternal inheritance pattern consistent with a mitochondrial DNA mutation. . . . The most severely affected patient had constant myoclonic jerking, dementia, ataxia, spasticity, hearing loss, and hypoventilation." There are probably a number of mitochondrial diseases that can now be identified and studied.

Although intracellular organelles cannot replicate outside their host cells, their degree of independence is best seen in the mitochondrial chromosome. In *Genetic Maps 1984*, volume 3, there is an impressively detailed map of those disturbingly independent parts of our existence.

The ability to map human mitochondrial DNA and to determine its genetic message in detail provides an opportunity to study human family relationships. This hereditary material has a high mutation rate, compared with chromosomal DNA, and many of its alterations appear to have little effect on fitness. These changes thus provide a whole series of experimentally accessible markers that we can use to trace family histories and ethnic and racial lineages. Since these lines of descent are strictly maternal, we don't

have to concern ourselves with the theoretical character
of reports on paternity.

The anthropological usefulness of mitochondrial DNA
studies is immediately apparent. Douglas Wallace of Emory
University has, in a personal communication, reported on
studies about founder effects in American Indian mitochondrial DNAs. In a study of southwestern American
Indians (primarily Pima and Papago tribes), he and his co-
workers uncovered typically Asian patterns of mitochondrial DNA but with dramatic differences in the frequency
of different markers. One rare Asian marker was found in
40 percent of the Indians studied. They were also able to
show significant tribal differences. The tentative conclusion is that "Indian tribes must have been founded by a
small number of Asian females." Doubtless there were
males along too, but we don't find out very much about
them in these studies.

The subject of human genealogy by mitochondrial DNA
is clearly in its infancy, but ethnology has been presented
with a new and very powerful tool, one that could hardly have been anticipated from classical studies of that
discipline.

This practice of musing on mitochondria, started by Dr.
Thomas, does have a certain fascination about it. The
thought of these once foreign invaders that are now an
essential part of all higher organisms does indeed induce a
strange feeling about us and our bodies. It now leads us to
do some second thinking about maleness and femaleness.
Mitochondria are a sort of congenital infection we get from
our mothers, but an infection absolutely necessary for life.
That pensée is going to take a long time to really sink in.

Pachyderms
and Peripatetics

E lephants, the largest land animals, are so huge in comparison to humans that it is not strange that they also loom large in our fiction, fantasy, and foolery. I suppose my favorite elephant joke comes from a college humor magazine of a more naive age.

> Freshman: I want to do something big, something clean.
> Sophomore: Go wash an elephant!

All of this thought about pachyderms comes from my great surprise at finding out that Aristotle once dissected an elephant. How come, you ask, am I getting involved in the business of an Athenian philosopher cutting up like this? The reason is more logical than it may seem at first glance. For perfectly sane, scholarly reasons, I was tracing the history of vivisection and came across a wonderful book entitled *Aristotle's Researches in Natural Science* (West, Newman and Co., London, 1912). Written by Thomas E. Lones, the work details the great savant's writings in the sciences and discusses the experimental bases of his findings. The elephant appears among the forty-nine species that scholars believe Aristotle personally worked on. Lones states at the end of his listing, "The inclusion of the elephant may cause surprise, but Aristotle's statements about it seem to justify its inclusion in the list."

The image of the great Peripatetic, elbow deep in the entrails of an *Elephas maxima*, should deal a blow to those effete teachers of philosophy who lack a blood-and-guts understanding about what the Athenian Academy was all about. Aristotle was first and foremost a biologist. Probably he was taught dissection by his father, an Aesculapian or priest-physician, who was bidden to impart that learning to his children. The Academy in Athens was a research institute where real plants and animals were worked on. There is a pernicious rumor, spread for the past five-hundred years or so by a bunch of nattering Neoplatonists, that Aristotle never did any experimental work. Lones's book effectively squelches that canard. Let's hear no more of it.

The next question is, where did Aristotle get his elephant? Dissection is one thing, but getting hold of the experimental material is quite another. Here we can only speculate, but the circumstantial evidence is pretty good. Alexander the Great was, of course, tutored by Aristotle, and when the student went off to conquer the known world, he is reported to have left funding for his teacher's institute. He also directed that interesting plant and animal specimens be sent back to the Academy by his troops.

Early in 326 B.C., when Alexander's army came down the valley of the Kabul River and crossed the Indus River, Raja Taxiles presented the conqueror gifts, including thirty elephants. As we know from later stories of Hannibal's elephants, those beasts traveled well, so sending a sample organism to Aristotle in Athens should have caused no great problem. Getting an elephant specimen for research suggests another story—one that is more easily authenticated because it was told to me personally by the investigator.

In 1962, I had the privilege of meeting Georg von Békésy of the Psycho-Acoustic Laboratory of Harvard University. Although a distinguished scientist and Nobel laureate, Dr. von Békésy was somewhat shy and a difficult man to converse with. Our first meeting was a few weeks after President Kennedy gave a dinner at the White House for all American Nobel prize winners. As a conversation opener I said, "I hear you had dinner at the White House." "Yes," he replied slowly, "the food wasn't very good."

The professor retired from Harvard a few years thereafter and opened a laboratory at the University of Hawaii. I occasionally met him on the campus during my sabbatical in Honolulu, and we talked about his new laboratory and his continuing work on cochlear mechanics and the functional organization of the human ear. Those conversations rarely lasted more than a few moments.

Some two years later I was back in Honolulu attending a rather formal banquet. I knew almost no one at the pre-dinner reception, and as I took my martini and stood alone, I noticed Georg von Békésy also standing alone, cocktail glass in hand. We exchanged pleasantries, and then the embarrassed lull in the conversation led each of us to a second martini. An announcement was later made that dinner was going to be delayed, so the two of us did the only reasonable thing. We each took a third martini.

By this time von Békésy's shyness was gone, and he became quite expansive and told me his elephant story. If I do not recall all of the details, you must realize that by this point in time I, too, was considerably relaxed.

Von Békésy was, in his early career, a researcher in Budapest. He had spent many years dissecting the inner ear of rat and human cadavers in order to establish the mode of interaction of sound transmission and sonic energy transduction in, as well as the anatomy of, the mammalian ear. Then one day the phone in his laboratory rang. It was the Vienna Zoo. One of their elephants had just died, and they heard von Békésy was in need of a fresh cochlear preparation. The scientist was off on the next train to Austria, and in one of the high points of his scientific career, he was able to use the greatly enlarged cochlea to help resolve some troublesome questions. He beamed while describing for me the dissection of an inner ear that was large enough to really get your hands on. I had never before heard anyone talk so passionately about the inner ear of an elephant. Indeed, I have never since heard such eloquence about otic matters.

That was the last time I ever saw von Békésy; he died a

few years later, and in the interim our paths had not crossed. Some time later while reading a review article by von Békésy on cochlear mechanics, I came across the following rather poignant lines:

> In the case of the elephant, of which I was very lucky to get a fresh cochlea, the basilar membrane is about 60 mm long, and the cochlea instead of being helical is coiled flat in one plane.

What is clear from all this is that those who need elephants for their research have to be patient. It takes something special like a world conquest or a death at the zoo.

Kraken

"Terrors of the deep" has been a recurring theme in literature at least since the Book of Jonah. Victor Hugo in *Toilers of the Sea* and Jules Verne in *Twenty Thousand Leagues Under the Sea* presented fearsome stories of attacks by giant sea creatures that rose out of the depths to engulf men and boats. Hollywood technicians have created their share of mechanical marine monsters in order to luridly portray those denizens of the depths.

Behind all this fancy and fiction lies some basis in fact, as I was reminded some time back by a photograph on the cover of *Science News* of a 450-pound squid that washed ashore on Plum Island, New York, in 1980. That specimen now rests in the Smithsonian Institution, where it has been the object of scientific study. Information about giant cephalopods had previously been assembled by Frank W. Lane in his fascinating book *Kingdom of the Octopus* (Sheridan House, 1960).

One of the problems in sorting out fact from fiction about large sea creatures is that encounters with them are rare, and those experiences have usually been sufficiently frightening that it is difficult to interpret the eyewitness accounts. Those tales, however, provided the only available early information. In Norwegian writings dating back to 1555, the giant forms were named kraken. In the first edition of *Systema naturae,* in 1735, Linnaeus included the kraken and classified them as *Sepia microcosmos.* His dropping of that species in successive editions reflected the skepticism and doubt about giant squid that existed among scientists.

Then, in 1861, a French navy dispatch boat, *Alector,* encountered and harpooned a "giant octopus" off the Canary Islands. The creature broke away, leaving a fragment of arm weighing about twenty kilograms. The spectacular battle was reported to the French Academy. The ship's officers estimated that the creature weighed 2,000 kilograms and reported that its mouth was shaped like a parrot's beak, opening fully half a meter. Canary Island fishermen had reported other sightings of squid so large that the fishermen were afraid to attack them. There was no longer any reason to doubt the existence of very large cephalopods.

The next specimen was brought in by three herring fishermen off the coast of Newfoundland. They had also done battle with a large squid and returned with a piece of tentacle 19 feet long and 3½ inches in diameter. All squid are decapods, having eight arms and two much longer tentacles. Octopuses have eight arms but no tentacles. A month after the first tentacle was recovered, another group of herring fishermen netted, killed, and brought in a squid that measured 32 feet from the top of its head to the tip of a tentacle. It was about the size of the animal pictured on the cover of *Science News*.

Several giant squid were sighted off the coast of Nova Scotia in the 1870s, and specimens were sought by the amateur naturalist Rev. Moses Harvey. Harvey contacted Yale zoology professor and mollusk expert Addison Verrill. That zoologist made a thorough study of the Newfoundland specimens and in 1879 published his classic papers in *Transactions of the Connecticut Academy of Arts and Sciences*, describing the giant squid and naming the species *Architeuthis harveyi* and *Architeuthis princeps*. Thus, these giants of the sea moved from myth into the domain of normative science. To honor Verrill's work, a full-size model of Architeuthis hangs in the Invertebrate Hall of Yale's Peabody Museum of Natural History. Suspended from the ceiling, it is a fascinating facsimile of the great creatures.

Establishing the reality of Architeuthis raises the question of the size of the largest giant squid. The issue is complicated by the small number of observations, probably related to the extreme depths at which the animals live. The only measurements reported are of squid that have risen to the surface or have been brought to the surface in the stomachs of giant sperm whales subsequently killed and cut apart.

In considering the size of great squid, two measurements are important: the length of the mantle, which is the body size, and the overall length from the end of the mantle to the tips of the tentacles. The largest complete specimen—beached at Lyall Bay, New Zealand, in 1888—had tentacles of 49 feet 3 inches and a body of 5 feet 7 inches. Various fragments that have been found suggest the possibility of 12-foot bodies and 70-foot overall lengths. Marks made by the sucker rims of squid on the skin of whales indicate that even larger specimens may exist, perhaps as large as 100 feet overall.

The chief natural enemy of the great squid is the equally titanic sperm whale. There are a few eyewitness reports from whalers and other seamen of battles between whales and squid. A full twenty-eight years before Verrill's scientific publications, Herman Melville wrote in *Moby-Dick* of "the great live squid." Indeed, in the course of the narrative a giant cephalopod is mistaken for the white whale, and the *Pequod*'s four smaller boats are lowered in pursuit of the animal. Melville was aware of what had been observed and written about the sperm whale:

At times, when closely pursued, he will disgorge what are supposed to be the detached arms of the squid; some of them thus exhibited exceeding twenty and thirty feet in length. They fancy that the monster to which these arms belonged ordinarily clings by them to the bed of the ocean; and that the sperm whale, unlike other species, is supplied with teeth in order to attack and tear it.

There seems some ground to imagine that the great Kraken of Bishop Pontopidan [the early Norwegian authority] may ultimately resolve itself into Squid. The manner in which the Bishop describes it, as alternately rising and sinking, with some other particulars he narrates, in all this the two correspond. But much abatement is necessary with respect to the incredible bulk he assigns it.

We still know very little about the natural habitat of Architeuthis. The habitat of the benthos has not been extensively explored, and there may well be many bottom-dwelling animals that have yet to find their way into zoologists' jars and pickling tanks.

I for one take pleasure in the existence of still unexplored portions of the world. They provide material for novelists and explorers and thus enliven the mundane with thoughts of the unknown. Clearly, one day there will be no unexplored places on this planet, and then our imaginations will have to reach outward into deep space and inward into the even deeper human psyche. Different terrors will accompany those searchers.

Societies

Breaking
the Law

R eports of the collapse of a bridge on Interstate 95 at Greenwich, Connecticut, set me thinking about death, disorder, and decay. Three very different pieces of literature came to mind. The first and most poignant was *The Bridge of San Luis Rey* by Thornton Wilder:

> On Friday noon, July the twentieth, 1714, the finest bridge in all Peru broke and precipitated five travellers into the gulf below. . . .
> Everyone was very deeply impressed, but only one person did anything about it, and that was Brother Juniper. . . . Anyone else would have said to himself with secret joy: "Within ten minutes myself . . . !" But it was another thought that visited Brother Juniper: "Why did this happen to *those* five?"

One hundred thousand vehicles had crossed the Greenwich bridge every day for a quarter century. The fate of the three people who were killed and the three who were critically injured makes it hard to avoid Brother Juniper's question.

But the issue of *why* the bridge collapsed, the mechanics of the disaster, sent me to that unlikely source of engineering knowledge, *Old Possum's Book of Practical Cats* by T. S. Eliot. There we read of a feline named Macavity, who has "broken every human law, he breaks the law of gravity." Mr. Eliot amuses us by playing on the two uses of the word "law": human law and law of nature. The first is broken daily; the second when broken ceases to be a law, according to philosopher of science Karl Popper.

Returning to Greenwich bridge's falling down, we see a clear demonstration of the law of gravity. But the reason for the disaster stems from humans trying to ignore another principle of nature: the second law of thermodynamics. That empirical generalization states that at the atomic and molecular level there is a tendency toward randomness and disorder. Since macroscopic structures are made of atoms and molecules, the effect leads to the breakdown of all ordered structures. To maintain structure it is necessary to perform work, that is, to expend energy in a directed way.

In the "wisdom" of the human body, a certain appreciable fraction of basal metabolism consists of energy used to repair structures that decay through protein denaturation, autolysis, and cell death. Much of the energy derived from what we eat goes into countering the disruptive tendencies of the second law.

There is no getting around the principles of thermodynamics. They are there, fundamental properties of collections of atoms and molecules. Technologies exist for reducing the rate of decay, but the dissolution itself is inexorable. To assume that it can be ignored for fiscal reasons is, in the words of a television commercial, trying "to fool Mother Nature."

The clue that the State of Connecticut was trying to ignore the second law of thermodynamics came when various officials and engineers were interviewed and someone let slip the terrible phrase "deferred maintenance." I must confess that I am hypersensitive to those two words; they have been a camouflage behind which I have watched the physical structures of several great universities decay. However, that reflects only my personal vantage point. I am told that most of the physical structures of the United States built in the past hundred years are also suffering from deferred maintenance. These include roads, bridges, railroads, dams, and buildings.

Attempts to defer needed maintenance are attempts to ignore or break the second law of thermodynamics. Clearly, this is not possible, so the only alternative is to burden the future for present benefits.

Individuals are often required by statute to maintain structures. Landlords are regularly fined for failure to maintain buildings, plumbing systems, and heating plants. Yet institutions and governments seem free to forget the need to counter the forces of decay.

In the case of commercial aircraft, regular maintenance without deferral is mandated by law (human). There is an absolutely fixed schedule of checks, repairs, and replacements that must be made. There are penalties for failure to comply. Perhaps the overwhelming influence of the law of gravity, in the case of airplanes, forces us to take the second law of thermodynamics more seriously. What is becoming apparent is that on a longer time scale we must take it equally seriously for less high-flying structures—or risk the consequences.

It is quite clear that deferred maintenance is more expensive in the end than regular renewal and rapid repair. Leaky roofs lead to rotted timbers, clogged drains cause flooding, unlubricated machinery grinds itself up, and worn wires cause fires. All of this is well known, yet the tendency to put off until tomorrow seems hard to resist.

Talk of deterioration suggests another poem. It is as if Percy Bysshe Shelley wrote a sonnet to deferred maintenance. He called it "Ozymandias":

I met a traveller from an antique land
Who said: Two vast and trunkless legs of stone
Stand in the desert . . . Near them, on the sand,
Half sunk, a shattered visage lies, whose frown,
And wrinkled lip, and sneer of cold command,
Tell that its sculptor well those passions read
Which yet survive, stamped on these lifeless things,
The hand that mocked them, and the heart that fed:
And on the pedestal these words appear:
'My name is Ozymandias, king of kings:
Look on my works, ye Mighty, and despair!'

Nothing beside remains. Round the decay
Of that colossal wreck, boundless and bare
The lone and level sands stretch far away.

If science poses the problem of decay, it also has the answer. The proper expenditure of energy can restore a system to its initial state. The next sound you hear will be all of us rolling up our sleeves.

Happy Holiday

W ill Rogers frequently pointed out that when he was low on humorous material, he would read the daily newspaper to find out what Congress was doing. It's getting so that the contents of my mailbox are becoming a similar source of amusement. Why, only today my credit card company wrote promoting some new accident insurance. Life insurance is, as we well all know, rarely considered humorous; for most of us it is a very direct dollars-and-cents confrontation with our mortality. However, let me describe the features of this coverage so you can judge for yourself.

First, there is $50,000 for accidental death, a benefit that requires little discussion. The second item is a $75,000 additional payment if the accident occurs while the insured is riding as a paying passenger "in a plane, taxi, bus, or train," later defined as a "licensed common carrier." These words certainly raise questions of definition. I recall the first long trip my wife and I took after we had a family. We purchased common-carrier insurance to provide for the tykes should the plane carrying the two of us not make its destination (one gets used to euphemisms in the insurance industry). Having arrived in Switzerland, we were riding up Mont Blanc in a swaying cable car (*téléphérique*). The magnificent scene took on a new aspect when my wife casually asked, "Is this a common carrier?" In any case, the present policy seems to exclude camel caravans, mule trains, and an unlicensed canoe I once took as a paying passenger on the Amazon.

The third benefit is an additional $50,000 if the life-taking accident occurs while one is outside the United States. That feature certainly should be anathema to our national recreation and travel industry. For if I held this insurance and wanted to learn sword swallowing or sky diving—both of which I've thought of and rejected—I would certainly go abroad. Extra protection to the tune of $50,000 is a powerful incentive to spend vacations outside the country. Why, with this foreign benefit and the $75,000 common-carrier feature, one might even be tempted to take a taxicab in Rome or Tokyo.

The phrase "outside of the United States" causes some difficulties for those who are involved in boating. Does the insurance company recognize the old three-mile limit, the conventional twelve-mile limit, or the new fisherman's 200-mile limit? How far out do I have to windsurf to make sure I'm protected by the extra coverage? In any case, it's clear that if the insured decides to go over Niagara Falls in a barrel, he should choose the Canadian side of that magnificent waterway.

The next item, and the one that prompted me to pen hese words, is called the $50,000 National Holiday Benefit. If the insured dies as the result of an accident on Thanksgiving, Christmas, New Year's Day, Memorial Day, Fourth of July, or Labor Day, then $50,000 is added to the basic cash payment. There it is, my friend; if you manage to have a fatal accident on a common carrier abroad on a U.S. national holiday, your grieving beneficiaries will be $225,000 richer. And who says there is no free lunch?

Actually, it is thinking about national holidays that really has me excited. Curiously, I find an enigmatic parenthetic section in the advertisement: "The specific number of days you're covered before and after each holiday is shown on your Certificate of Insurance." Good Heavens, credit card company, do I have to actually pay for the insurance before you'll tell me how soon before and after the holidays I have to plan my dangerous activities?

And then I worry about a change of national habits if this kind of coverage becomes widespread. Those of us who stay home on New Year's Eve for fear of drunk drivers will now be out on the highways, undaunted and backed up by a full 50 thou of additional coverage. I expect that come the first day of the year, half the citizens of San Diego will be bombing around Tijuana, Mexico, backed up by a whole hundred thousand dollars of additional support.

And come Thanksgiving, will we be sitting around stuffing ourselves on the traditional turkey? Well, perhaps, but on the other hand it seems like the perfect time to go to Japan to try the delicious and dangerous fugu (pufferfish), preferably eaten in the dining car of a moving train. For the Fourth of July there will be the urge to go to Taiwan, where the biggest and loudest fireworks are still available. I envision driving around Taipei, lighting giant crackers and throwing them out of the window of a moving bus, within which I am, of course, a paying passenger. I leave each reader's imagination to guide him or her to the thrills available through the new innovation provided by one of America's leading insurance companies.

By now, suspicious chap that I am, it seems natural to be reading the fine print. After being told "YOU CAN'T BE TURNED DOWN FOR COVERAGE" in capital letters, then 7-point type continues, "coverage when you're 80, 90— even 100 years." Gracious, if one is going to live to be 100, why the need for this kind of insurance? But wait—also in those tiny letters is the footnote, "Rates increase at age 65." A little quick mathematics reveals that the increase will be a full 61 percent. Come, come, you actuaries up there in Hartford, don't you realize that after 65 I'm going to give up skydiving and race-car competition, take fewer ski tows, and sell my motorcycle? Do you really think the odds will be 61 percent greater that I'll be done in on a foreign common carrier on a U.S. national holiday? It's hard to know how you chaps figure.

And speaking of Hartford, the fine print also states, "Coverage is not available to persons age 65 and over in Connecticut." Does that mean I'm going to have to look forward to commuting seventy-five miles every day just to retain my coverage? I feel that, as a minority, the senior citizens of Connecticut are being selected for discriminatory treatment. If this is group coverage, as you say, let's not start out in violation of the Civil Rights Act or perhaps the Constitution itself. (*Secretary, please make a photocopy of this article for the ACLU.*)

And while I'm on the civil rights issue, I have an objection to your Husband-Wife Plan. It, as you know, covers the cardmember for $50,000 and the spouse for $25,000. This is going to create a lot of dissension in families about who holds the card. And do you really think such an individual's life is worth twice that of the spouse? I know it's an honor to carry your card, but does that make me twice as valuable as my mate? No, you may be good at finances, but you have a lot to learn about married people and equal rights.

And so, credit card company, if you don't get my application back, do appreciate all the thought and angst that went into this carefully considered decision.

Digs

Recently I became an archaeologist. It was not a planned career change, rather something that just happened. My university responsibilities had shifted to include supervision of a residential college, and I found myself peering into a basement room full of storage items left behind when students had departed. Abandonment had not been the intent, but, alas, storerooms are often repositories of good intentions displaced by the flow of events. I recall that at our laboratory we periodically clean the freezer of UFOs (unidentified frozen objects). These are tubes and vials put away in anticipation of further experiments that never materialize. And so it was in the basement.

What does one do with a treasure trove of artifacts from American college life of the last dozen years? First, there was an attempt to locate the owners. This proved frustrating; for when found, they often could not remember having deposited the items, denied their existence, or disavowed any interest in these strange reminders of a not-too-distant past. The archaeological studies were interfacing with the anthropology of a consumer culture,which discards possessions in a rather casual manner. What surprised me most were the cartons of books. As I picked up copies of *Chaucer's Poetry* and the *Basic Writings of Aristotle* I could not help thinking.

> *A Clerk ther was of Oxenford also. . . .*
> *For him was levere have at his beddes heed*
> *Twenty bookes, clad in blak or reed,*
> *Of Aristotle and his philosophye.*

This society is far from the ways of the threadbare scholar, whose few possessions were his treasured books.

Having attempted to locate the owners, we moved to step two, an auction of the abandoned material. This gave an opportunity to move ahead with archaeologists' work, cataloguing materials. In order of frequency of occurrence, they were books, articles of clothing, trunks and suitcases, sports equipment, furniture, electronic devices, typewriters, and art items (paintings, sketches, prints, sculptures, and the like).

Most of the books were texts, but the variety of recreational reading was astounding: novels, poetry, how-to manuals on sex, cooking, and running. There were even a few collector's items, including handsome leather-bound volumes.

Sports equipment tended to be seasonal, such as skis, ice skates, and snow boots. Even in this purely objective study, I was beginning to develop thoughts about the parental sweat that had gone into earning the substantial capital that was now being consigned to the highly depreciated auction block. The furniture tended to chairs and end tables, and was all reusable. The clothing seemed quite serviceable.

In the electronics category we came to those objects that most uniquely characterize the age we are studying. Amplifiers, tuners, speakers, television sets, and radios were in greatest abundance. Most of these were in working condition. Some already seemed quaint, as obsolescence comes very quickly in rapidly developing technologies. Very few vacuum tubes have survived into the microchip age.

The typewriters, surprisingly, included electric as well as manual models. They constituted a focus of life for paper-writing students, but they, too, were abandoned. The art included the most personal of items, but in our brief study we found it impossible to further classify these artifacts.

After the auction many of the objects were strewn about, unsalable. The best of these were taken by a custodial employee to be distributed at her church. The residue, which could be neither sold nor given away, was set out

for trash collection. On the street, seemingly out of nowhere, there materialized a group of trash sorters; recy-
clers, I choose to call them, who carted away some of the
remains. The residuum left in a dump truck and was, I
assume, more thoroughly recycled in the city incinerator.

I am left with the raw data, trying to formulate theories
about the kind of society that would leave such a reposi-
tory. If my find had been 2,000 years old and I had no other
information about the culture, I would have concluded that
the inhabitants left their possessions in haste under some
severe threat, such as war, famine, or pestilence. I would
have eliminated war as a possibility, because the victors
would surely have looted such a valuable store of materials. Even pestilence would have left survivors to start again
with these possessions. No, I would have concluded that
famine drove these people to migrate, and because of death
or traveling too far, they never returned. I would have
shed a tear or two for this destroyed civilization.

Of course, my archaeological postulates would have
been totally wrong. The inhabitants of these buildings were
well-fed, healthy individuals, living in a period of peace.
They had ample time to gather their possessions together
and ample facilities to ship them elsewhere. These items
were abandoned by choice, not by chance.

It is generally acknowledged that we live in a society
guided by consumerism and deeply influenced by a throwaway policy. My archaeological investigations alerted me
to the full extent of that ethic among the young,who no
doubt are the strongest exponents of this viewpoint. Our
economy is geared to the throwaway, and a real return to
thrift and conservation would lead to massive unemploy-
ment and social disaster. We are producing artifacts faster
than any previous culture; getting rid of plastics and old
automobiles is a major problem. A trip to the town dump
is a fascinating study of our life and times.

Given these economic arguments, why do I commit the
archaeological sin of getting judgmental about the culture

I am studying? Why am I troubled about that basement full of abandoned possessions?

The worry comes from the thought that we are renouncing a dimension of the past in our lives, in our relation to our possessions, in our relation to our culture, and in our relation to other people. Life is lived so fully in the present that there is little thought for yesterday's stereo and, perhaps, for yesterday's memories.

The second concern stems from my feeling for Thoreau's doctrine: "That man is the richest whose pleasures are the cheapest." The individuals who abandoned the basement items seem to have quite expensive pleasures, particularly when they have to be bought and rebought each time one moves.

I do not have an answer, but I am comforted by looking at a carpenter's plane that lasted my grandfather a lifetime or a cooking pot that my great-grandmother brought to the United States from Poland.

Xictmd and Quetzalcoatl

I recently glimpsed the future, and frankly, I'm feeling a little shaky. The occasion was a conference held at the Xerox International Center for Training and Management Development at Leesburg, Virginia. I was there not as a Xerox trainee but was attending a conference being held in space rented from Xerox.

We arrived by car via the beautiful rolling hills of northern Virginia. A long road into the center went past a guardhouse, where we were asked for a good deal of information about the nature of our business.

The center consists of tiered pyramidal ceramic buildings suggestive of Maya or Aztec temples. At the registration area we parked briefly and got our maps, room assignments, and instructions, as well as a brief lecture on the room-numbering system. Each room is assigned a four-digit "codon." The first digit denotes the level, ranging from 1 to 4. The second indicates the module, and the last two designate the room. Thus, "4310" directs one to the fourth level, third module, tenth room. In addition, the modules are color coded: blue for 1, yellow for 2, orange for 3, red for 4, and violet for 5. The rooms are not numbered consecutively, and the classrooms seem to be organized on a different system from the residence rooms. The decor is modern American penitentiary style. I sat down on the bed and thought, "Welcome to Stalag Xerox."

Except for the first floor, it was not possible to go from module to module without changing levels. The halls were filled with people wandering about, lost. I began wandering about, trying to get a feel for the place, but that intuition never came. I, too, was lost.

Food was served in the dining room, Module 3 (orange) Commons. Off the dining room is a cafeteria called a "student servery." The ambience of the dining room can best be conveyed by the following instruction: "The cafeteria service in the student dining room is a 'shopping center' concept. You should not form conventional cafeteria lines. Instead, make your selection of items at random, moving about the area and selecting from those stations (salad, hot entrees, dessert) which are least crowded. With large numbers of people in the serving area at any given time, short lines can form momentarily, but they should be at right angles to the serving counter rather than parallel to it. In the dining room, we ask that you eat from your own tray [Please, may I use a knife and fork?] so that when you are finished you can carry your tray and dishes to a soiled-dish window, located at each side of the dining room, adjacent to the condiment stand."

The facility also has a snack bar and cocktail lounge, miscellaneous game areas for pool and Ping-Pong, and, in another building, a fitness and recreational center. Trails through the woods are provided for joggers and walkers.

One other house rule should be mentioned—the dress code. Again, it is best quoted: "In keeping with the Xerox policy, we do ask that blue jeans or athletic attire be worn *only* after 5:00 P.M. Slacks and sport shirts are acceptable during the normal business hours (8:00 A.M.–5:00 P.M.)."

The impact of the center on most of our conference attendees was immediate and strong. Coming mainly from universities and government, they were quite unfamiliar with this kind of ambience. I do not know whether I was viewing the mind of corporate America or examining the vision of one individual, but it seems worthwhile to explore some of the features that contributed to the psychological impact of the setting.

First and foremost is its total isolation. Most of the participants in our conference came by bus from Dulles Airport. At the center, there is no public transportation and no place to go nearby. All of one's life must take place within the confines and rules of the "campus."

Second, the uniformity and starkness of the decor keep saying "institution, institution, institution." The center has a uniform, mechanical feel. Work space, meeting space, and living space are completely mixed together, so there is no escape from the central busyness. Perhaps a beehive is a better analogy than a prison. Indeed, the center seems designed for social insects, a frightening view of a future in which that model has been chosen for human society.

The modular architecture itself reinforces the insect analogue when viewed from the inside. From the outside, the impression persists of a Maya or Aztec temple. Walking around the exterior of the building, one keeps waiting for an announcement of when the virgins will be sacrificed to Quetzalcoatl.

The numbering and color-coding system clearly dehumanize the habitat and contribute nothing to efficiency. People seem to spend a great deal of time wandering around lost. Most of the committee meetings I attended started late because participants were trying to find the conference rooms. Some people told me they felt like experimental rats in a maze.

On my second night at the center, having nothing to do and tiring of the cocktail lounge, too noisy for decent conversation, I resolved to wander through all five modules and master the layout regardless of how long it took.

Frequently, in getting from here to there, one exited a module onto a plaza and then located an entrance to the next module from that plaza or another, one flight of stairs up or down. At each such doorway from the module to the outside, I noticed on the ground a small gray metal box, about the size of a one-gallon carton, chained and padlocked to the doorpost. There was some black printing on each box, but alas, my bifocals focus poorly at that distance. Finding one doorway with no one around, I got on my hands and knees and read, "Rodenticide." The vision was ominous: not only were we struggling through a rat's maze, but the punishment was tangible and well defined. Of course, there was nothing strange about rat poison in a building in the woods. Nevertheless, the location and labeling were all consistent with the disquieting aura of the center.

As a sometime reader of grim futuristic novels, I don't quite know what to make of Camp Xerox. It is clearly one view of the future; it would make a fine set for a dream scene in a Woody Allen movie or a real scene in a *2001*-type drama. *Brave New World* could also be filmed in those surroundings if the cocktail lounge served soma and the beds were more "pneumatic."

I left the meeting seeking some insight into the mindset of those who had planned this edifice. What corporate mentality had decreed this kind of place? What architect had chosen such spaces for human beings? What manager had created the numbering system? And who had decided to enclose the people inside of rodenticide-baited doors? I do not know, but I whistled "Dixie" very loudly as my cab passed through the guarded gate and I was once again out on Route 7.

The Sport
of Eggheads

I recall a number of summers ago meeting an acquaintance I had not seen for several months. He greeted me with the news, "All the intellectuals are taking up golf these days." The only available reply was, "How long have you been playing golf?"

Well, now that the president of the American League is a distinguished physician and the late president of the National League was a former university president, the time has come to acknowledge baseball as the sport of intellectuals. If I had any doubts about that idea, they were resolved last summer when I inquired after the health of a venerable aging metaphysician and epistemologist. The reply was, "As long as the Red Sox keep winning, he's in good spirits."

To establish personal credentials as a baseball fan, I must tell of expeditions that went by train from Poughkeepsie to Manhattan and then by subway to Ebbets Field in Brooklyn. Those were the days of "the boys of summer," when watching the Dodgers play bordered on the mystical. In particular, the games between the Dodgers and the Giants evoked a feeling akin to a religious experience.

Actually, my days of fandom predate the trips to Ebbets Field, for Poughkeepsie had a semipro team that played on Friday nights against strange groups of nines that traveled around the country challenging local clubs. They had unusual names, but I can recall only the "House of David," which consisted entirely of bearded players. The weekly enterprise was supported by passing the hat some time around the third inning. Apparently, the take among the thousand or so fans was large enough to keep the contests going. The manager of the Poughkeepsie team lived in the same two-family house as did my grandparents, and as a youngster I felt I knew a real celebrity. Visiting teams often included special individuals intended to attract crowds. Two of them stand out in my memory. One was a one-armed right fielder who so impressed me with the power of will over adversity that I can still see him catching a fly ball, getting his glove off, getting the ball into his hand, and throwing it to the infield. Another team had as its first-base coach the legendary Grover Cleveland Alexander, who won thirty or more games a year for three consecutive seasons (1915 to 1917). Alas, Alexander was hitting the bottle in those days and had to be replaced by a pinch-coach in the fifth inning. People in the stands muttered about what a shameful example it was for the children in the crowd. I was about seven years old at the time and was excited just to see any one of the "immortals," drunk or sober.

Among other all-time greats I have watched on the field was Satchel (Leroy Robert) Paige. It was the summer of 1952. We were living in Kensington, Maryland, and my next-door neighbor suggested that we go to a night game between the St. Louis Browns and the Washington Senators. Washington baseball is remembered by the old adage, "First in war, first in peace, and last in the American League." In 1952, the Browns and Senators were probably in contention for last place, a fact that does not deter the true fan.

It was a good game, and in the eighth inning Paige was brought in as a relief pitcher by the Browns. Satchel was probably forty-six years old at the time, though scholars are uncertain as to his exact age. The game was tied, Paige and his opponent pitched flawlessly, and the crowd thinned as we went through the 10th, 11th, 12th, 13th, and 14th innings. I can't remember when the game ended or who won, but no matter: I had seen the great man perform in the twilight of his career and in the early morning of my workday.

It is impossible to think about Satchel Paige's achievements without wondering how blacks could have been excluded from professional organized baseball for the first three score and ten years of that endeavor. From the perspective of today, it is hard to face the fact that we are only 125 years away from a slave-holding society and only 40 years from an exclusionary national sport. Those are unconscionable flaws in a society otherwise dedicated to opportunity for all its citizens.

Speaking of flaws, there is a historical scandal associated with the history of baseball. In 1908, the Spalding Commission reported that the sport was invented by Abner Doubleday in Cooperstown, New York, in 1839. The *Encyclopaedia Britannica* goes to lengths to demonstrate that baseball was not invented but evolved from the English children's game of rounders and that the Spalding Commission's report was largely a fiction created to demonstrate that baseball was of American origin. Another burden of the intellectuals is to straighten out the history of their favorite sport.

Regardless of its origins, baseball grew to maturity in the United States and was for many years a uniquely American institution. Its popularity has now spread to Latin America and Japan.

Professional baseball was certainly unknown in England in the immediate post–World War II period. I recall taking an English couple to Ebbets Field in 1950 to show them an authentic bit of Americana. The four umpires came out in their formal black suits and small black caps. They lined up together for the playing of the national anthem. The Englishwoman turned to me and inquired, "Who are the gentlemen in mourning?"

All of the preceding is background to the question of why baseball is so popular among professors and philosophers. Why is fanaticism—the act of being a baseball fan, or fanatic—so common in academic circles? This involves two separate lines of inquiry: 1) Why do people become fans of professional sports teams? and 2) Why do intellectuals become baseball fans?

The first of those queries is so deep, so psychoanalytic, and so Jungian in its impact that we cannot address it in this brief essay. Identifying with warriors goes very far back in human history, and we will for present purposes simply take the phenomenon as a given. Each city, of course, has its own kind of fans. In Boston they are a curious mixture of little old ladies from Pawtucket, Rhode Island, Jesuit novitiates, M.I.T. professors, blue bloods, blue collars, and bluestockings, with a generous measure of hard hats and high techs. Such is the institution of fandom.

The appeal of baseball to intellectuals stems from the fact that the sport is so individualistic, so complex, so four-dimensional, so statistical, and so meditative. Baseball is a team sport, but with a very important difference. The action at any instant involves one to four individuals very intently carrying out highly skilled tasks. Most of the eighteen players are largely spectators while the acts of daring take place. Each fielder must, however, always be alert, should the action shift toward his territory. Each individual must thus be considered—his on-field mannerisms, his idiosyncracies, his imperfections. Baseball involves a philosophically inviting balance between the individual and the team.

The complexity of baseball stems from a simple set of operations that may be performed in a large array of permutations and combinations. As a result, the rules are elaborate, and four officials are required to be in constant attendance. To sense the complexity, note that the renowned Casey Stengel invented an entire grammar and vocabulary to deal with the subtleties of the sport. He felt that the philological and semantic difficulties lay outside the scope of English syntax and took the decisive step of linguistic innovation.

The four-dimensionality stems from the shape of the field and the pace of the game. Football, laid out on a rectangular grid, tends to be two-dimensional. Baseball, because of the irregular shape of the field and the height to which the ball is hit, clearly utilizes all three spatial dimensions. Because it is an untimed game, unlimited by any total number of points, it has its own time dimension. Thus, it has a certain affinity to a relativistic age. No location compares with the centerfield bleachers of the Oakland Coliseum for sitting and pondering the true meaning of Einstein's work.

The statistical, indeed the actuarial nature of the game of baseball is another unique feature. In no other sport do spectators sit, pencil and scorecard in hand, poised to record each and every action. I can recall students flunking arithmetic who could work out batting averages to four significant figures. In no other sport is the official scorer so important a player in the unfolding drama as fans, eyes fixed on the scoreboard, wait to see whether an H or E flashes.

And because the sport is so actuarial, every move a player makes becomes part of his permanent record. One is reminded of Omar Khayyám's

> *The Moving Finger writes; and, having writ,*
> *Moves on: nor all your Piety nor Wit*
> *Shall lure it back to cancel half a Line,*
> *Nor all your tears wash out a word of it.*

As a highly statistical sport, baseball is a perfect match for the modern age of computers.

The meditative nature of baseball also flows from the pace and principle of local action. Most players spend most of their time contemplating what the future will bring. Aside from the pitcher, catcher, and batter, all are silent, waiting and preparing their souls for the unknown. It is this constant pre-action meditation that relates baseball to metaphysics and Zen and joins it to the perennial philosophy. Small wonder that one of the leading players was nicknamed "Yogi."

And as I turn on the television set, I reiterate: "All the intellectuals are baseball fans these days."

Nameless
Terrors

D espite extraordinary advances during the current age of science, the human psyche, individually and collectively, is still haunted by terrors that from time to time disturb our sleep or bedevil our waking hours. When we arise in a cold nocturnal panic or experience diurnal sweaty palms, terrors that probably originated in the deep primordial forest sometimes come into consciousness. Psychologists have adopted the term "phobias" to describe morbid fears. In more recent times, the word has tended to be used to describe irrational states of mind, though in a world where perception and reality are so intertwined, it is difficult to disentangle rational from irrational sources of alarm. One is reminded of the therapist telling the patient, "No, you don't have an inferiority complex, you *are* inferior."

The most fearsome terrors are the nameless and faceless ones, the inchoate entities that go bump in the night. Medical science has long been aware of the special dread of the nameless, and much of early science has gone into designation, using the classic elegance of Greek or Latin, of each and every complex of symptoms. This is the first step along the long road toward a rational therapy. Hence, we have generated voluminous medical dictionaries, a mellifluous prose genre.

I first became aware of the importance of providing a name for undesignated terrors many years ago when I heard the following story. A physician on rounds was surprised to see a critically ill patient sitting up, eating, talking, and remarkably improved. He inquired about the wondrous cure. The patient explained: "Last night when I was lying here half conscious, I heard you talking to the other doctors. You gave the diagnosis with such certainty that I just knew I would be all right." The attending physician, who could not recall exactly what he had said, summoned the resident, who did remember. "You must have assumed the patient was unconscious, for in a louder and more assertive voice than usual, you said to me, 'This man is moribund.' "

In dealing with people in all walks of life today, I have noticed a new nameless phobia so persistent and so widespread that it has become a dominant feature on the American scene. Whether one talks to physicians, teachers, custodians, or business people, concern over getting sued is pervasive, so much so that many innovative and imaginative activities, as well as some very practical ones, are grinding to a halt. A kind of national paralysis is setting in, driven by this use of the law. The syndrome is clear; the etiology is not. Are we suffering from a morbid, irrational fear of lawsuits or from the rational fear of a legal system gone awry?

In the tradition of the centuries, I decided that the first step in dealing with the new mania is to assign a name to the condition. Then we can proceed toward its cure. After all, how can one write a grant application for an unnamed disease? The first candidate for the novel word was "litigophobia," but this combination of a Latin beginning and Greek ending is offensive to scholars. I took the matter directly to my colleagues in Classics, and a day later the proper term emerged: "dikaiophobia." With two Greek roots (*dikai* means lawsuit), it satisfies the scholars and gives us a proper term for fear of being sued. (Special thanks are due to Professor George P. Goold of the Yale University Classics Department.)

Having formulated a proper name, we can proceed to the substantive questions. Is dikaiophobia a psychological reaction such as agoraphobia (fear of open places), or is it a physiologic sensitivity such as photophobia? Both of the last two negative reactions drive people out of the sunlight, but the psychodynamics are very different.

What of dikaiophobia? It is difficult to classify with the highly individual fears such as agoraphobia. If it is nevertheless a morbid condition, it must be grouped with social pathologies that sometimes seize a whole segment of society and propel it along strange paths.

A classic example of such an epidemic of the psyche is tarantism, a form of hysteria that existed in Italy from the fifteenth to seventeenth centuries. It was supposedly caused by the bite of the tarantula and was cured by frenzied dancing that would sometimes obsess entire communities. Vestiges of this madness persist today in the tarantella, a lively, flirtatious dance.

Victims of tarantism believed they were responding to real spider bites. If dikaiophobia is not a pathologic response to "imaginary spider bites" but is a sane response to something in our legal system, then we are part of a society gone mad. Dikaiophobia is a social madness, either the insanity of groups of individuals or a madness incorporated in the very fabric of our legal system.

At this point, there is a great temptation to launch into an attack on lawyers, an activity that would doubtless strike sympathetic chords in many readers of these pages. And a few months ago, I would have been tempted to so respond. In the meantime, however, I have thought seriously about an essay by Charles L. Black, Jr., in the book *The Humane Imagination.* Black, an eminent authority on the Constitution, advises lawyers that when they are assaulted at cocktail parties with the Shakespearean quote "Let's kill all the lawyers," they should respond by reciting slowly the names Sir Thomas More, Abraham Lincoln, Mohandas K. Gandhi. After a discussion of justice and injustice, he reminds us, "Law does change for the better, and in a short period sometimes, and always through the energy of lawyers." Alas, Charles Black has finessed my ability to take cheap shots at lawyers.

Reflecting further on the roots of this problem, I realized that we are a society that has handed over to lawyers the solution of conflicts. This leads to all the obvious abuses we are so well aware of. Having given lawyers such awesome powers, we can hardly fault them for vigorously representing their clients. The fault, dear Brutus, is not in our lawyers, but in ourselves.

A second negative component of dikaiophobia is its enormous "cop-out" value. Individuals in all walks of life can hold back from interesting, bold, and innovative actions by affirming, "That would subject us to the danger of lawsuits." Institutions and corporations also hide behind the threat of legal action.

I suspect that beyond all the other reasons for dikaiophobia is a certain avariciousness and excessive concern with self that most of us find hard to resist. Given the legal avenues to pursue get-rich-quick schemes, so many people sue that there is a real danger that anyone could be capriciously dragged into court. Wrongful suit has few risks and penalties.

There are clearly long- and short-term cures for the terror we are discussing. The first and obvious one is to pursue better relations between individuals, an elusive goal we always seek. Over the short term, certain changes in the legal system could substantially reduce the number of questionable lawsuits and clear the air enough to allow a more rational approach to righting wrongs.

A few years ago, a parent of a very troubled college student threatened me with a $10 million lawsuit because I had persuaded his child to seek psychiatric help. On reflection, I suppose that in these times it sometimes takes the threat of a suit to persuade one that he is doing his job with adequate vigor.

Suds

A book entitled *Encyclopedia of Surface Active Agents,
Volume II* is not ordinarily calculated to send one off
into a discourse about social theory, but Table II did catch
my eye. The table is headed "Per Capita Consumption of
Soap and Synthetic Detergents" (as reported from 34 countries in an international survey of production for 1953–
1954). The final column is labeled "Domestic Computed Per
Capita Use Soap and Surfactants (Pounds)." How, I ask,
could anyone's imagination fail to be titillated by this
numerical assessment of a whole complex of national
habits?

Top among the sudsy set are the Australians at 27.9
pounds annually for every man, woman, and child. Second
in the cleanliness derby is the United States at 26.6 pounds
per capita, and the United Kingdom is third, coming in at
26.6 pounds per head. Clearly, the first postulate is that
speaking English and using soap are correlated. The
strength of this linguistic conclusion is rather weakened by
the second triad on the soap scale—Switzerland at 24.6,
Israel at 24.5, and The Netherlands at 24.3 pounds per person. We seem to be skipping around the world with little
geographic or geopolitical rationale. At least the figures reinforce my childhood image of the Dutch housewife in
wooden shoes scrubbing the front porch.

Not wishing to be deterred by the failure of the first
hypothesis and realizing that the social sciences are inex-
act, I will move on to Belgium and Luxembourg, Denmark,
and West Germany with rates of 23.3, 22.4, and 21.9
pounds per capita. The next cluster contains New Zealand
at 21.4, Sweden at 20.3, and Canada at 19.5 pounds per
person. The differences are now becoming significant, with
7.1 pounds a year separating the United States and Canada.

This is not to make dirty remarks about our neighbors to the north. After all, Canada is a far colder country, and sweat must be a major factor in frequency of bathing. When a final theory of national cleanliness is produced, it will have to account for mean national temperature and humidity as well as the total use of washing aids.

Uruguay at 17.0, Cuba at 16.8, and the Union of South Africa at 16.6 take us to new parts of the world, while France at 16.6 and Portugal at 15.6 return us to Europe. Regrettably, many nations are missing from this study, so our conclusions are going to be tentative, at best. For example, there are no data given for Eastern European countries, so we remain in the dark about the correlation of soap and socialism, or cleanliness and communism, as the case may be.

The figures we do have lead us to ask why the Belgians and Luxembourgers used 40 percent more soap and detergent every year than their more southerly neighbors, the French. This goes counter to our temperature theory; since parts of France are warmer, the effective cleanliness difference by our reasoning above is even greater. We lack the theoretical foundations to account for the fact that the French are world leaders in producing perfumes and colognes, a possible substitute for soap in human social relations.

Next on our list are Norway at 15.4, Austria at 14.7, Ireland (Eire) at 10.9, and Italy at 10.5. The next to last number comes as a surprise when compared with the United Kingdom's value of 26.2, a value that I presume includes Northern Ireland. It is hard to think of the cultural differences between those neighboring lands that would lead one to do two and one half times as much washing and cleaning as the other. Travelers to Eire find the country clean, green, and pleasant. It is often hard to know what to make of these sudsy statistics. Indeed, one feels that a whole new social discipline could emerge from the studies and we are only touching the surface.

The next eight values for pounds of soap consumption per person per year are listed in order: Jamaica 10.3, Colombia 9.6, Philippines 9.5, Panama 9.1, Brazil 8.6, Guatemala 7.7, Venezuela 7.3, and the Dominican Republic 6.6. The countries are mainly South and Central American and the consumption reflects the life-styles and economic conditions in that part of the world.

As one gets to the bottom of the list, a few caveats are in order. Most countries that consume low per-capita amounts of washing materials do not have reliable usage statistics, so that the data become very fragmentary—some one hundred countries are missing from the list. In addition, the data are thirty years old and I suspect a current cleanliness survey would show a somewhat different set of figures. Nevertheless, the table at hand concludes with Japan at 6.2, Egypt at 5.8, Peru at 3.2, and India at 0.8 pounds per capita. One wonders, for example, if communal bathing helps to keep the Japanese use of soap at a rather low level for such a technologically advanced country.

The range of values is enormous, the top consumer using thirty-five times as much material as the bottom user. In an effort to make some sense out of these numbers, I followed a hunch and went to life-span tables for various countries. Out came the graph paper, and the widely varying numbers were put in order by plotting life span against soap consumption. From 0 to 12 pounds per person per year, the life span (for women) rises linearly from 26 years to 72 years; above 12 pounds a year, life span is unrelated to use of washing material. The initial increase of 3.83 years of life per pound of soap used is a powerful argument for technology, or washing behind the ears. My mother would have appreciated the argument when she urged a recalcitrant five-year-old into the bathtub with the words, "Cleanliness is next to godliness." It is as I have always suspected: While soap in small quantities is good for you, above a certain level it is a matter of culture, social mores, and advertising. Not a bad result when one stops to think about it. For developing countries, soap consumption is a measure of wealth and health resources. For developed countries, the uses of soap are much more aesthetic and symbolic—the Norwegians, at 15.4 pounds per year, are outliving by a few years the Australians, who are using 80 percent more suds. That interesting, if enigmatic, fact is probably a good place at which an amateur sociologist should stop and turn the problem over to professionals.

The Grate Society

It was a rather cold morning, the day was clear, and the Washington Monument stood forth in the sunlight. As I walked down 21st Street toward Constitution Avenue, I was getting that satisfying, grade-school-indoctrinated "I am an American" feeling that often hits me when I visit the nation's capital. My euphoria was jolted when I saw, near the intersection with Virginia Avenue, a group of men sleeping on metal grates that were the exhaust outlet of some underground heating system.

It was not the first time I had seen people sleeping on heating outlets; that loss of innocence occurred in Berkeley on the University of California campus many years ago. Those sleepers were hippies, or drugged-out survivors of the flower children, or other deliberate dropouts from the mainstream. They seemed to be going through a temporary phase in their lives, but I could not predict its duration. Although their presence was disturbing, it seemed like a local Berkeley phenomenon.

My second encounter with street dwellers had a greater impact. I was attending a conference in New Delhi in mid-winter and encountered large numbers of people huddled in rags—sleeping on the ground or living in groups clustered around sooty fires. So numerous were their fires that the smoke formed an acrid miasma, which hung over the city as a constant reminder of life on the streets. In India one grows to think that those problems have been around for centuries, even for millennia; thus one's sensitivity is dulled.

However, seeing street people and the Washington Monument in the same view was simply overwhelming. Here was a stark montage of our failure to fulfill the dream of our Founding Fathers. And those lying before me were, I knew, but a small sample of the more than a quarter of a million homeless Americans. This was not the capital of India, with its age-old problem, but the capital of the United States, with a newly generated national disgrace.

It was half an hour before my meeting, and I continued down to Constitution Avenue, with its magnificent buildings celebrating all aspects of national life. Without much cerebral involvement, my feet led me to the Lincoln Memorial, which acts as something of a magnet, attracting those who wish a place to stand and ponder the national dream. After a few moments in the presence of the Great Emancipator, I was drawn to the Vietnam Memorial. To those of you who have not visited this shrine, I can describe the scene, but I am not sure I can convey the sense of what was happening.

Across from the Lincoln Memorial, in the park by the reflecting pool, a simple walkway gradually leads down to a large depression about six or seven feet below ground level. On the steep side of this declivity, acting as a retaining wall, is a continuous series of black marble slabs, on which are engraved the names of the 53,000 Americans who died in the Vietnam war. It is not a monument to a horrible war, but a memorial to individual human beings sacrificed for some unclear purpose.

Next to the wall, families and friends of the deceased place flowers, American flags, and newspaper articles. The visitors find names they are looking for, finger the names in the marble, take pictures, make stone rubbings. There is enormous personal interaction between the visitors and the monument. Some leave letters in the cracks between the marble slabs.

At Jerusalem's Wailing Wall, petitioners leave letters to God in the spaces between the ancient stones. At America's wailing wall, lovers, parents, children, and friends leave letters to the deceased. Because they departed so quickly, so precipitously, so unexpectedly, there were many words left unsaid. And so the survivors set down these words on paper and leave them in the wall, somehow establishing a bond between themselves and the fallen.

It is a very heavy emotional experience one undergoes in visiting this national shrine. And I left shaken. As a country, we have been cosmically unfair to those who were of military age during the slaughter. To those who accepted service, we denied the thanks a nation owes individuals who risk their lives for their country. To those who refused service, we denied the thanks a people owes those who take unpopular and costly stands out of conscience. To those who returned, we denied the love necessary to overcome the horror of what they experienced. And to those who did not return, we denied life itself.

Coming back through the park to Constitution Avenue, I mused that the Vietnam war and the homeless people on 21st Street are not unrelated. Somewhere along the way we wavered in our sense of national purpose. We forgot that the dignity of each individual is part of that goal. If in the midst of plenty we allow homeless, friendless men and women to live and die on the streets, then we individually and collectively fail to uphold that compact made in Philadelphia to "insure domestic tranquillity" and "promote the general welfare."

Columnists and essayists are constantly accused of being sermonizers, and I have made a great effort over the years to avoid preaching whenever I find that genre slipping into my writing. But my morning walk in Washington somehow so silenced my sense of humor that I simply have to share with you the feeling that we cannot ignore our social problems, lest Washington in twenty or fifty years become indistinguishable from New Delhi. There have been only twenty-five years from Lyndon Johnson's Great Society to today's Grate Society. Yet, surely there are no insoluble social problems, only failures of the will.

The Vietnam war so distracted us from our sense of national pride and desire to succeed that we have been drifting aimlessly, replacing real accomplishment with the bureaucratically orchestrated illusion of accomplishment. I want to assure you that the Grate Society is real enough, and from what sociologists tell us, it is growing rapidly. Walk through your own city and look.

I do not know how to finish this essay, for the work is so ill-defined. I can only think of the words of the Talmudist Rabbi Tarfon: "It is not thy duty to finish the work, but thou are not at liberty to neglect it."

The Computer
Revolution

Computers
and Brains

As the debate about the relation of computers to the human brain unfolds, it seems surprising that no one among the diverse participants has paid much attention to relevant ideas expressed thirty and forty years ago by two of the wisest thinkers of our century, Erwin Schrödinger and John Von Neumann; for the thoughts of those men have much to say to the current discussion.

Schrödinger was one of the founders of quantum mechanics, and the celebrated equation bearing his name is still the starting point of many applications of that science. In 1944 he turned his thoughts to biology and wrote a wondrous book called *What Is Life?* That thin volume influenced a large number of physicists who in the postwar period discovered biology and its many challenges to the worldview of physical scientists.

The chief question in *What Is Life?* is, how do living cells manage to achieve such precision of genetic operation in the face of the randomness of thermal noise at the molecular level? Schrödinger starts by pointing out that most of the laws of physics show such a high level of precision because they involve averages over vast numbers of molecules. Since fluctuations are random, the averages are exact. This type of behavior is designated as "order from disorder," a phrase that describes the mode of operation of all ordinary machines.

When one starts to miniaturize machines into the micron range, the fluctuations due to thermal effects (Brownian motion, for example) interfere with the smooth operation of these systems, and orderly functioning becomes impossible. The smaller the machine gets, the more serious the noise problem becomes. How then, asked Schrödinger, could genetic self-replicating machines operating at the level of single molecules manage to maintain such fidelity for thousands and millions of years?

To answer the genetic question, the quantum physicist discussed "order from order machines," molecular devices that are dependent on covalent bonds, which are highly energetic in comparison with thermal energy. These are, in effect, quantum mechanical machines, which operate in a radically different manner than do "order from disorder" machines. The distinction drawn by Schrödinger is a very important one. There are two radically different ways of building precision equipment: macroscopic and macromolecular.

In spite of impressive advances in miniaturization, all computer components used to date have operated in the "order from disorder" domain. Cellular components involved in brain function may conceivably function in the "order from order" domain, as in the case of genetics. We must hold out the possibility that computers and brains use radically different hardware. Knowing whether or not this is so can come only from a detailed study of the molecular biology of the central nervous system. Until this issue is resolved we must at least be conscious of the possibility that a major difference between brain and computer is due to their operating with hardware that functions in different physical domains.

John Von Neumann was one of the pioneers in the development of high speed computers. A broad-ranging mathematician and physicist, he was at the center of a group at the Institute for Advanced Study that in 1946 developed JONIAC, a first-generation computing machine. For heuristic aid in developing programming he began to study neurobiology. His ideas on the relation of computers and brains were sufficiently developed that he accepted an invitation to give the Yale University Silliman Lectures on the subject in the spring of 1956. These plans were thwarted by a severe illness. When Von Neumann died early in 1957 he left behind an uncompleted manuscript, which was published in 1958 by the Yale University Press under the title *The Computer and the Brain.* The essay is enigmatic because the author seems to have envisioned a radically innovative view of brain function that his illness prevented him from being able to formulate completely. In spite of that incompleteness his ideas are well worth reviewing.

The hookup of neurons through synapses and the all-or-none character of the action potential dictate that the brain, according to Von Neumann, must act as a digital device. That is, in so far as the brain computes, it must be of the class of digital rather than of analog computers.

Computer elements were 10,000 times faster than the brain in 1956, and I suspect that advantage has gone up a couple of orders of magnitude or more with semiconductors. When Von Neumann compared the elements of biologic networks to computer elements with respect to size, he found electronic devices 100 million times larger than naturally occurring components. The use of semiconductors has, I am sure, narrowed that discrepancy by a good deal. In the final area of comparison the computer elements have a much higher precision than their living counterparts.

When all of these hardware differences are considered, Von Neumann concluded, the brain must operate with a logic that avoids the necessity of the arithmetic precision of the computer. He wrote:

The Language of the Brain Is Not the Language of Mathematics
. . . whatever language the central nervous system is using, it is characterized by less logical and arithmetical depth than what we are normally used to. . . . The statistical character of the message system used in the arithmetics of the central nervous system and its low precision also indicate that the degeneration of precision, described earlier, cannot proceed very far in the message systems involved. Consequently, there exist here different logical structures from the ones we are

ordinarily used to in logics and mathematics. They are, as pointed out before, characterized by less logical and arithmetical depth than we are used to under otherwise similar circumstances. Thus logics and mathematics in the central nervous system, when viewed as languages, must structurally be essentially different from those languages to which our common experience refers.

. . . whatever the system is, it cannot fail to differ considerably from what we consciously and explicitly consider as mathematics.

On that radical note the book abruptly ends, terminated by the author's untimely death. He was challenging the validity of the underlying conceptualizations we use to study the brain and compare it with computers. Yet what is surprising, given the great esteem for John Von Neumann, is that no one has really taken up on his argument and fully developed its consequences in the mind versus artificial intelligence arguments that have been waging the last few years. To be sure, the great mathematician was ahead of his time, but thirty years have now passed since his ideas were formulated.

What Is Life? and *The Computer and the Brain* are both small, profound books, written by men of genius yet accessible to a broad scientific reading public. Although they may seem out of date, they both probed such fundamental issues that the ideas they presented continue to reverberate. Both books were written by theoreticians who saw in the hardware of life profoundly different structures from those employed in contemporary machines.

There is a tendency among scientists to assume that ideas from thirty and forty years ago have either been incorporated into the paradigm or are no longer relevant. In the case in which no consensus has emerged, that assumption is emphatically not true. The writings of our very best thinkers possess a kind of timelessness if they are not lost on dusty library shelves. Those interested in comparing human and machine behavior would be well served by a study of these two classics.

Information,
Please

The year 1949 marked the appearance in the *Bell System Technical Journal* of a series of papers entitled "The Mathematical Theory of Information." That truly seminal work by C. E. Shannon and Warren Weaver engendered great excitement, and in the following years a variety of publications applied information theory to everything from the works of Shakespeare to noisy transmission lines. The original articles were highly mathematical but relied at the core on the relatively simple notion of defining the information content of a message as the logarithm of the reciprocal of the probability of receiving that message in fact. The more unlikely it is that the message will be received, the more information can be gained from its receipt. That new and easily understood concept spread with surprising speed through the fields of electrical engineering, psychology, physics, biology, and at least a dozen other disciplines.

The communications industry, which has always placed a premium on transmitting the maximum information per unit cost while keeping the transmission as error-free as possible, was the first to find major applications. After all, telegraph companies have been sued when damages resulted from the receipt of a message other than the one that had been sent.

Although it is generally recognized that Shannon and Weaver were the first to define measurement of information with mathematical clarity, the idea seems to have been in the air during World War II and in the immediate postwar period. Norbert Wiener, the father of cybernetics, generalized the mathematics he invented to help shoot down airplanes and pondered the relation of information and entropy. Alan Turing, the mathematical genius who had devised methods of breaking the German Enigma code and saving British shipping from submarines in 1940, had introduced the logarithm of the ratio of two probabilities as the "weight of the evidence." In 1943 the American economist A. Wald had used the logarithm of probability ratios to construct "plausibility," a concept he employed for quality control of goods from factories during the war.

In a curious way the great technological war forced people to choose between alternatives, and a metric was needed for how much one's knowledge is increased by a message or a measurement or by any other factor that would change the uncertain probabilities of alternatives. The logarithm of the probability, or its negative, seemed to have emerged as a natural measure.

The basic postulate of information theory, however, predates World War II. In the quieter days of the mid-1930s, statistician H. Jeffreys published a paper entitled "Further Significance Tests." He had used the logarithm of the ratio of likelihoods and had given the name "support" to that function.

The publication of Shannon's paper and the subsequent discussions of information and entropy reminded some physicists of a 1929 paper by Leo Szilard, in which he had to introduce an information measure similar to that eventually introduced by Shannon in order to show the implications of the second law of thermodynamics at the molecular level. Szilard later achieved fame as the man who persuaded Einstein to inform Roosevelt of the potential of the atomic bomb. In the 1920s and '30s he was engaged in fundamental problems of physics. The logarithm of the probability measure emerged from Szilard's work in a very intuitive way by comparison with the entropy function, which also involves a logarithm of a probability.

But we have not yet completed our trip back into time. Our story, insofar as I can trace it, begins with Charles

Sanders Peirce (1839–1914). He was one of the idiosyncratic group of mid-nineteenth-century Americans who "did it their way." It is a group in which I count Melville, Gibbs, Poe, Whitman, and Thoreau. Yet among those luminaries, *Encyclopaedia Britannica* recognizes Peirce in a special way. It says that he is "now recognized as the most original and most versatile intellect the Americas have so far produced."

Charles was the son of a Harvard professor and a graduate of that prestigious college. His formal scientific career was largely spent with the U.S. Coast and Geodetic Survey. From 1891 to 1914 he was a free-lance intellectual living in relative isolation, in his own words a "bucolic logician." He had earlier founded the philosophy of pragmatism, cofounded semiotic psychology, contributed to linear algebra, logic, philology, geodesy, cartography, meteorology, and astronomy. He also seems to have conceived an information measure seventy years before Shannon's paper.

In 1878 there appeared in *Popular Science Monthly* an essay entitled "The Probability of Induction" by C. S. Peirce. It dealt, among other things, with intensity of belief and the author's feeling that there ought to be a measure for the intensity of our belief in a scientific proposition, a thermometer on which "our belief ought to be proportional to the weight of the evidence." He then went on to argue that "the *logarithm* of the chance" is the best measure of belief. This Peirce measure is, of course, just the negative of the Shannon measure of information: The more strongly we believe something, the less information we get from a confirmatory message.

Peirce went further and suggested that the reason for the logarithmic form might be related to the nature of the central nervous system. He wrote:

There is a general law of sensibility, called Fechner's psychophysical law. It is that the intensity of any sensation is proportional to the logarithm of the external force which produces it. It is entirely in harmony with this law that the feeling of belief should be as the logarithm of the chance, this latter being the expression of the state of facts which produces the belief.

It was a bold proposition and one that might be profitably reexamined in somewhat different terms today. For the notion of information is not an abstract statement about the world out there but rather a statement about the human mind and how we evaluate input. That is one of the reasons Shannon's work has been taken so seriously by psychologists.

In any case, the Peirce measure of the degree of belief is the negative of the Shannon information measure and was introduced for very similar reasons. In that sense Peirce was clearly a forerunner of information theory.

We are left with the intriguing question of why an idea that was barely noticed in 1878 was hailed, acclaimed, and widely used in 1949. Science abounds with examples of premature ideas that were first ignored before being applauded; I suppose Gregor Mendel's paper on the genetics of peas is the most widely known example. But Peirce was not an unknown; he was a member of the National Academy of Sciences and a fellow of the American Academy of Arts and Sciences. Another 1878 paper of his in *Popular Science Monthly,* "How to Make Our Ideas Clear," is considered by many to represent the beginnings of the American school of pragmatic philosophy. Although Peirce was clearly a loner, his works did not go unread.

While the whole notion of premature ideas is an important one in the sociology of science, it is not clear what makes an idea ripe for widespread acceptance.

From Cabala
to Entropy

I must confess to being a chronic bookshelf scanner. Upon walking into someone's office, it's just impossible to resist noting the books and drawing conclusions. In bookstores it's a matter of stumbling through aisles, trying to notice as many titles as possible. In reference sections of libraries the habit becomes most severe and causes the greatest difficulty, for each book of a multivolumed series records a strange mélange of alphabetical associations.

Thus, one passes *Collier's Encyclopedia,* and "Amen to Artillery" sets off a stream of consciousness. *Encyclopaedia Britannica* offers "Metamorphic to New Jersey," which seems like the beginning of some great cosmic joke. *Encyclopaedia of Religion and Ethics* gives us "Sacrifice to Sudra," which triggers all sorts of thoughts of East and West. One of my prize specimens of this literary genre is "Trance to Venial Sin" from *The Encyclopedia Americana.* But without doubt my all-time favorite is from volume 2 of the 1967 edition of *The Encyclopedia of Philosophy.* Its spine reads "Cabala to Entropy."

What is so intriguing about these verbal dyads is that they lead on, defying us to seek some relation between the two members, to make some sense of a pair of alphabetically related but otherwise apparently unconnected concepts. And so one sits in the reference reading rooms pondering cabala and entropy. How can such disparate constructs possibly enhance or explain each other? The answer is some time in coming.

Cabala (or kabbala) is esoteric Jewish mysticism that dates back to who knows when and lays claim to who knows what secret knowledge of God. Cabalists have been concerned with the *En-Sof* ("the Infinite"), the most remote and hidden, unapproachable aspect of Godhead.

Entropy is a much newer word invented by Rudolf Clausius in 1865 from the Greek ἐν ("in") and τροπή ("turning"). It was introduced as a measure of the portion of a system's energy unavailable for conversion to mechanical work. The concept of entropy rapidly developed into one of the central ideas of the discipline of thermodynamics. Yet most physical scientists found entropy a very difficult notion to grasp, one that required much study, thought, and time. Well, it's a long way from *En-Sof* to entropy, but have faith, dear reader.

Somewhere between the third and sixth century of our era, cabalists devised the notion that the creation of the universe involved ten divine emanations and the twenty-two letters of the Hebrew alphabet. As the tradition continued, there developed a letter and number symbolism of great variety and complexity. By the time the theme had run its first course, every word and letter of the Pentateuch had been given a mystical explanation.

That theoretical cabala led to practical or applied cabala involving magic formulas and attempts to manipulate the world with letter and number magic. Thus, what began as the search for the inner nature of divinity on occasion deteriorated into a mumbo jumbo of not-understood symbols inscribed on amulets. Cabala moved from religious awe to numerical prestidigitation. And so into the modern world.

Returning to entropy, we note that Rudolf Clausius's work followed from that of Sadi Carnot and was motivated in large part by a desire to increase the efficiency of steam engines or at least to determine what limited the efficiency. Entropy has always been important to engineers, especially those mechanical engineers whose professional activities were closely geared to the thermodynamic steam tables.

A transition from practical to theoretical entropy occurred in the works of Ludwig Boltzmann and Josiah Gibbs, who developed methods of understanding entropy in terms of the atomic and molecular states that underlie the large-scale observed properties of all matter. Their work led to the science of statistical mechanics, which allows one to calculate observed physical properties by averaging them over all the chaotic random behavior at the microscopic and submicroscopic levels.

And so things stood until the end of World War II and the formulation of information theory by Claude Shannon and others. A feature of that information measure was that it was represented by almost exactly the same formula as used for entropy in statistical mechanics, with the probability of a system being in some atomic state replaced by the probability that a given message is sent or received. If you're wondering what this had to do with cabala, I must beg your continued patience.

Because information is central to every area of human endeavor, the idea spread from the communications industry where it began into the worlds of psychology, aesthetics, the social sciences, and the humanities. With information theory came the nebulous concept of entropy that had been wrestled with by generations of physicists and engineers. A group of theorists outside of the physical sciences became intrigued with the notion of entropy. Letter and number symbolism arose using entropy-related formulas borrowed from physics. Because the practitioners of this new approach were undisciplined by the hard reality of building steam engines or doing experiments, their use of parts of the formalism of physics sometimes also became a letters and numbers magic. Those who then tried to use this material to influence the world around them were engaged in a kind of practical cabala.

Thus, concepts of entropy and cabala both led, in the hands of some extremists, to attempts to manipulate events through the use of abstractions far removed from the empirical constraints of everyday life.

Well, we've done it. We've closed the circle between

entropy and cabala. The result shows the possibilities of interrelating diverse ideas. It also shows the dangers inherent in attempting to give too much meaning to such symbolic arrays as the guide words on the spines of encyclopedia volumes. With that in mind, I'll refrain from actually crafting the joke from "Metamorphic to New Jersey," even though I think it has great potential.

Dr. Mitzkin's Mistake

I had just settled back for some recreational reading. After all, the short stories of Isaac Bashevis Singer are sufficiently removed from biological thermodynamics as to provide a mini-vacation from day-to-day professional interests, or so I thought until I came to the ninth page of a tale with the fascinating title "A Friend of Kafka." There, in clear black and white, were the sentences

> In his old age he was learning to dance, influenced by the philosophy of his friend Dr. Mitzkin, the author of "The Entropy of Reason." In this book Dr. Mitzkin attempted to prove that the human intellect is bankrupt and that true wisdom can only be reached through passion.

And in the next paragraph the thought continues:

> He came to practice what he preached, but he could just as well have written. "The Entropy of Passion."

A thought that had been lurking in my mind for some time suddenly came into focus: There is a syndrome that I shall designate as entropophilia, love of the word "entropy," which is endemic among contemporary writers, whatever their subject matter may be. Authors adore that word and delight in using it for its own sake. Isaac Bashevis Singer, master teller of tales of ghosts, disembodied souls, and demons, was subtly getting involved with that scientific terror, Maxwell's demon.

Entropophilia on the part of an author like Singer was puzzling because it was so unexpected, but there was also something else troubling in this little vignette about Jacques Kohn, the man who knew Kafka. I went back to the beginning of the story and in the first paragraph confirmed that the time and place were "the early thirties . . . in Warsaw." There was the problem: an anachronism in the setting of the story. *The Entropy of Reason* would have been a perfectly plausible concept in New York in the 1960s, which is where and when I imagine this particular story was written, but it was a most improbable title in 1933 Warsaw.

The word *"entropie"* was coined by Rudolf Clausius in Berlin in 1854. It comes from the Greek and means a transformation. Entropy in classical thermodynamic usage is a numerical quantity describing the state of a system. It can be measured with precision only at equilibrium. Entropy is one of those quantities that are immensely useful in physics and engineering and at the same time very difficult to understand in any easily visualizable way. With the advent of kinetic theory and statistical mechanics, at the turn of the century, the idea developed that the entropy value measured the degree of randomness with which energy was distributed among the atoms and molecules making up the material being studied.

For the first four decades of the twentieth century entropy remained a highly technical term known primarily to physicists, chemists, and chemical engineers. Thermodynamicists were involved with working out the many problems made accessible by the works of Josiah Willard Gibbs. It was an epoch of detailed scientific work in thermal physics, and it is very hard to imagine that by 1933 the idea of entropy could have moved far enough into the domain of social thought to suggest that it be linked with reason or passion in a single title.

In 1948, the publication of *The Mathematical Theory of Communication* by Claude Shannon caused entropy to move from engineering handbooks into popular writing, since the Shannon measure of the average information content of a message bears a striking resemblance to the entropy formula in the statistical mechanics of Gibbs. Indeed, one can reinterpret entropy as being proportional to the average information we would have if we knew all the microscopic detail about the state of the system, that is, if the positions and velocities of all the atoms were known. Since such detail is generally lacking, entropy measures the missing information, or our ignorance of a system's exact state.

Information theory was applied to a wide array of disciplines, and with those applications the concept of entropy found its way into the humanities and social sciences. We had books on entropy and art, entropy and economics, and a variety of other areas. In this information-theory context, Dr. Mitzkin's *Entropy and Reason* was a striking example of entropophilia. It seems a bit harsh to criticize Singer for this technically rooted anachronism; we shall rather reserve our wrath for those who, like Dr. Mitzkin, indulged in and continue to indulge in the promiscuous use of "entropy."

Entropy, as introduced by Clausius, had a very precise meaning, definable by an equation and measurable by calorimetry. It had the units of energy divided by temperature and could be measured only for those equilibrium systems where temperature had a clear, unambiguous meaning. When the quantity was redefined for statistical mechanics by Ludwig Boltzmann and Josiah Willard Gibbs, they took great pains to show the identity between the entropy of statistical mechanics and that of thermodynamics.

Those who later tried to take the terminology of thermodynamics over into the humanities and social sciences often lacked a deep appreciation of the entropy concept in physics. As a result, entropy became a fashionable code word used to cover uncertain, vague ideas with a cloak of respectability. The difficulty came with the assumption that the various concepts introduced had properties analogous to those of entropy in physics. Thus the second law of thermodynamics was also co-opted and misused in cases where the flow of energy kept the system far away from equilibrium.

Curiously enough, those who use "entropy" in a loose fashion never seem to form concepts in their own fields analogous to temperature. This seems strange because entropy and temperature are reciprocal quantities in the thermodynamics of Clausius and his successors. Energy terms in classical theory have a component involving both entropy and temperature. If entropy in its "other-culture" sense is in any way to be analogous to its use in thermal physics, then something corresponding to temperature must be found. I believe this is avoided by entropophiles because temperature is well enough understood by those outside science to make clear the difficulties of any analogy. The formulators chose, rather, the obscure and difficult entropy concept to work with rather than the reciprocal, but much better known and intuitively understood, idea of temperature. In essence, if you don't understand something and want to sound profound, use the word "entropy." If that's what Dr. Mitzkin was doing, he was at least correct in worrying about the bankruptcy of the human intellect. I can't say much about *The Entropy of Passion,* but some of us get pretty passionate about misuses of "entropy."

From Botanical
Gardens to
Libraries

For me a trip to Sweden would have been incomplete
without an opportunity to visit the garden of Carl
von Linné (Linnaeus), the great classifier whose system
of biological nomenclature still casts a giant shadow on the
landscape of biology. One of the simpler joys of travel is to
stand on the ground where our great intellectual forebears
once trod and try to sense the thoughts and feelings they
had as they ventured into uncharted realms. The scientist's
house and small plot of land in Uppsala did not disappoint
me. Its neat array of plantings resonated with the orderliness of a mind of such scope and precision as to include all of the living world of his day in a coherent
scheme.

To try to place ourselves in Linné's cultural framework
we go back to the early 1700s, over a full century before
evolutionary thought replaced the permanence of Adam's
naming and Noah's saving of existing species. Each type of
animal and plant was then a representative of a divinely
ordained group or species. To recognize a plant or animal's
membership in a species, biologists had to have one or more
standard examples (type representatives) to which the test
organism could be compared. And so the great zoos and
botanical gardens of the world were organized to keep on
file species archetypes, platonic ideals of biological taxa.
Species could also be identified by graphic (later, photographic) and by verbal descriptions. The descriptive criteria
served well for giraffes and elephants, but they were much

less effective for grasses, mosses, and many other taxa whose speciation depended on subtle differences. For those examples, collections of organisms were a must.

With the development of microbiology, descriptions of organisms became more abstract, relying on microscopic and biochemical criteria. For strains of bacteria and other asexual forms, the question of what constituted a species became even more obscure, and the necessity of clones of archetypes became vital. They were at first maintained by individual laboratories and finally merged in collections, such as the American Type Culture Collection. Most strains could be freeze-dried and cold stored for long periods, but others had to be kept going by serial transfer.

Through all of the period from Linné to around 1900, two parallel methods of species identification were in use: type specimens for comparison, and ever more detailed descriptions for identification and keying of unknowns. A parallel system existed in organic chemistry, whereby one could identify a compound by comparison with a known sample or by checking it against a detailed description in various compendiums. As chemistry became better understood and the instrumentation became more precise, samples were often totally replaced. In subtle cases, such as pH measurement, comparison is still made to samples provided or by companies or government standards institutes. In general, however, the better a material is understood, the easier it is to maintain its description in a library than keep samples.

With the development of genetics around the turn of the century the problem of maintaining stocks became even greater. Each wild type species could give rise to any number of mutants at each locus, and (in principle) each mutant had to be kept as a standard for subsequent experiments. Storing a variety of corn seed is relatively easy, but maintaining pure lines of 10,000 or more different fruit flies is an enterprise of considerable magnitude.

Next, the rise of tissue and cell culture created a need

for maintaining those lines. Thus, the initial zoos and botanical gardens expanded to a rather diverse collection of organisms, cells, and other biologic materials.

Molecular biology is now giving us a new view of collections. The starting hypothesis is that all genetic information is stored in the sequence of nucleotides in DNA; the information is transcribed to RNA and then translated into protein; and the proteins determine an organism's activities. If that hypothesis is correct, then storing the complete sequence of an organism's DNA is the informational equivalent of maintaining the type organism. The shrub in Linné's garden could theoretically be replaced by a floppy disk giving its DNA sequence—as would equally be true of elephants, giraffes, and *E. coli.*

Therefore, to make an elephant, one would proceed in the following way. Go to the library and check out the floppy disk that says Elephant, Asian (*Elephas maximus*). Read out the disk and carry out the synthesis of the DNA molecules listed on the disk. Take a fertilized elephant egg, produced by any couple of elephants or from a nonspecific egg bank, remove the DNA, and inject the newly synthesized DNA. Implant the egg in a surrogate mother, wait two years, and you will have a juvenile copy of the archetypical elephant.

The experiment proposed above is premature and very iffy, since it 1) assumes no cytoplasmic inheritance, 2) makes no special assumption about the physical form of the DNA, and 3) assumes a mother cell will be available for the DNA insertion. The first assumption is wrong for organisms, such as elephants, whose DNA contains such organelles as mitochondria and chloroplasts. The floppy disk would have to contain additional information on cytoplasmic inheritance. The second and third assumptions about the DNA's environment represent currently unresolved experimental questions. Producing an elephant from a library is a long way off, but there seem to be no major conceptual barriers to its accomplishment.

The procedures for going from the library to the organism would be similar for bacteria and elephants. In fact, with the tiny one-celled organisms we are much closer to meeting the three assumptions and therefore have a higher probability of success. Bacteria have no organelles with independent coding, and the physical form of their DNA seems much simpler than the elaborate chromosome structure and accompanying mitotic apparatus of higher organisms. We have available techniques to sequence an entire genome of a bacterial size. I suspect that this will be accomplished within the next five years for the smallest DNAs. With bacteria we do not require a surrogate mother. The most difficult problem at the moment is obtaining a nonspecific host cell with its DNA removed—again, an experimental issue that requires a certain amount of trial and error.

In any case, we are at the beginning of an era of moving from botanical gardens, zoos, and type collections to libraries, computer retrieval systems, and synthesis of organisms from information. Make no mistake about it, this is a radical idea in human intellectual development. If we were to get good enough at this fast enough, we would not have to worry about endangered species; they could simply be kept on file in the library. Somehow I have the feeling that old Carl von Linné would have liked that idea.

The Whole
and the Sum
of the Parts

"The whole is greater than the sum of the parts" is recalled by many of us from an introductory biology course, where the expression was used to convey the idea that the integrated activities of a cell or organism present features that are not necessarily evident in the properties of the organelles or organs that make up the larger structures.

While the maxim was thought by some to be a bit esoteric though deeply meaningful a half century ago, the overall concept of wholeness can now be understood in terms of systems theory and network analysis. The extraordinary use of feedback control in living systems makes possible a high degree of stability, and global systems properties are an important part of the current paradigm. These ideas, introduced in 1948 by Norbert Wiener in his book *Cybernetics,* have been demonstrated experimentally at many levels of organization.

In recent years, with the rise of informatics, the relationship between the whole and the parts in biology has taken on another meaning. There is conjecture that lurking within the vast collection of experimental data (the parts), there are integrative features that will become new wholes. These principles are the theoretical content of biology. Because the number of data is so vast that they must be stored in computerized banks, the feeling exists that the search for integrative principles will be aided by techniques of artificial intelligence.

Two classical examples can be cited in which the proper management of data collections led to deep insights. For Carl von Linné (Linnaeus), the primary data of biology were the descriptions of the multitude of plants and animals that had been discovered and described by generations of observers all the way back to the time of Aristotle. This was a diverse and not very coherent collection of data. Among the problems in organizing it was the lack of agreement on how to define a species; there is still some trouble with this question. The sources Linnaeus dealt with were of many different types—correspondence, journal articles, books—and they reflected varying reasons for naming and describing species. The founder of the modern scheme of classification imposed on this collection of taxa a hierarchical method of grouping species into genera and then into higher categories, which gave rise to the following:

Kingdom

Phylum

Class

Order

Family

Genus

Species

This scheme can be considered an information-management device that led to the arrangement of a very large body of data into a coherent whole. The notion of a taxonomic tree was inherent in the hierarchical scheme.

When Darwin developed and published his ideas on the origin of species, the tree structure of Linnaeus's *Systema Naturae* stood as one of the data bases on which to ground the idea of common descent and speciation. Although Linnaeus believed in the special creation of each species, his principle of data management was so powerful that it was of enormous value to the evolutionists, who held quite a different view. In the new interpretation the vertical dimension of the classification tree became the time axis of evolutionary theory.

Some one hundred years after Linnaeus developed the protocol to organize the data of biology, Dmitri Ivanovich Mendeleev, of the University of St. Petersburg, struggled to make sense of another vast body of data. First the alchemists and then the chemists discovered a small number of elements and a large number of compounds. A substantial and diverse literature described the properties and reactions of these substances. The data bank was large and confusing and required an organizational principle. Mendeleev discovered that by placing the elements in a rectangular array and arranging them in order of increasing atomic weight, certain regularities occurred. It was later discovered that organization by atomic numbers gave a better fit, but the very notion of atomic numbers required the prior existence of the periodic table as set forth by Mendeleev. The periodicity gave rise to the octet rule, and ultimately the entire periodic table of the elements was explained by quantum mechanics and the Pauli exclusion principle.

Post–World War II developments and the enormous growth of molecular biology once again present us with a vast array of data, now larger by many orders of magnitude than anything faced in the past. Once again we must confront the problem of managing data to gain insight into deeper underlying principles despite great diversity. A number of biologists believe that hidden within the data are several integrative principles that will provide understanding of signaling strategies and developmental protocols. Certain intertaxonomic overlaps in hormones, signaling peptides, and genome segments suggest that there are important biologic laws remaining to be discovered.

Because the data bases are so large and the information range so extensive, it is difficult to envision anyone holding it all in mind to seek general relations. Rather, the problem suggests using expert systems and artificial intelligence programs to process the data in order to seek their connections and interactions. Biology, being in part a historical science, is different in structure from physics. Biologic theory to date has been less powerful and less general. In physics, a small number of axioms generate a vast range of observable relations. Biology has depended on local theories more restrictive in their applications. The use of computer data banks and artificial intelligences suggests new ways of creating a theoretical structure for biology. This novel approach is going to require a new breed of theoretical biologist, an individual trained in both information science and experimental biology.

Data are being generated with great rapidity in biology. The number of entries in the 1985 *Biological Abstracts* is 110,004. They are not isolated observations but part of a coherent structure generated by the laws of physics and chemistry, the common line of descent, and possibly as yet undiscovered biogenic principles. As Linnaeus and Mendeleev have so eloquently demonstrated, the appropriate rules of data management can be invaluable in transforming mountains of data into gems of insight. Fortunately, the great surge of new data has been accompanied by the growth of an information science and computer technology adequate to the range of material. Our task now is to put information science and biology together to develop a deeper understanding of the nature of life.

Past, Present, Future

It is a quiet Saturday afternoon, and I sit in the library of St. John's College in Santa Fe, New Mexico. This institution has a curriculum based upon the "Great Books," those hundred or so classics that are judged to have been most influential in molding contemporary thought. The educational format lays heavy stress on the dialogue across the generations, the importance of making education an integrative experience over the range of human civilization.

My reason for being at this place at this time is to participate in a workshop on computer data bases in biology and the use of computers, data management techniques, and artificial intelligence approaches in developing theories for biology and medicine. For the next few weeks, two large on-campus laboratories provide facilities for small computers, larger work stations, and telephone hookups to major data banks throughout the country. I am conscious of being at a way station between a past that reaches back to the oral tradition of Homer and a future in which machine-assisted thought will be generating truly novel approaches to human understanding.

As I query data bases in distant cities, two distinct thoughts come to mind. One is the idea of Marshall McLuhan that, "Today, after more than a century of electric technology, we have extended our central nervous system itself in a global embrace, abolishing space and time as far as our planet is concerned." The other is the notion of Teilhard de Chardin that the evolution of reflective thought has produced the noosphere, a network of interacting thought that encircles the globe and promises to change the planet as profoundly as the advent of the biosphere billions of years ago.

I am trying to find a thread that winds its way from Thucydides to Turing, from Vergil to von Neumann, from Machiavelli to Minsky. The task is a difficult one. It is always possible to look back and assess a historical development, but living in the middle of a great change, how does one evaluate it?

As the mind begins to drift, I wonder who among the greats of the past would be comfortable amidst us as we stroke our keyboards and peer at our screens. Who could step out of the past, don a pair of blue jeans and a T-shirt, and join us on the second floor of Evans Science Building?

Clearly, the first one to consider is Aristotle. For of the 1,425 pages of extant writings of the philosopher appearing in two volumes of the Britannica Great Books Series, 338 pages are, in fact, a data base of biology, and another 95 deal with human biology. Their data-base character stands out as one scans the titles: *History of Animals, On the Parts of Animals, On the Motion of Animals, On the Gait of Animals, On the Generation of Animals*. Down to the very last details, Aristotle tries to create a complete catalogue of all knowledge of animals. Surely he would warm to our idea of a great knowledge base drawing on all the data bases of biology.

Hippocrates, the contemporary of Socrates, would certainly be a candidate for our workshop. He collected and organized the knowledge of medicine of his time. His books (*Prognostic, Epidemics, On the Articulation*, and others) were sources of data from which he reasoned to formulate the rational management of patients. He would have welcomed any assistance that could have been called up from MEDLINE or generated by any of the diagnostic software programs. Galen would have joined his predecessor in this search for diagnostic clues.

I do not know if Euclid, Archimedes, Apollonius, and Nicomachus would have come to our workshop, but if they had, I'm sure they would have joined the artificial intelligence work group and engaged in the many discussions of how to structure data bases so as to derive the maximum knowledge from them. They, perhaps, would have been unsympathetic with our concentration on data and would have searched for more general principles. I suspect that these analytical mathematicians would have grown impatient with computer modeling.

Titus Lucretius Carus, whose interest covered an encyclopedic range, would, I imagine, have entertained himself scanning abstracts on a vast range of topics. He would have reveled in the retrieval programs and found vast realms of material to incorporate in a new version of *On the Nature of Things*. He might now, perhaps, be overwhelmed by the range of material he tried to include in his long poems.

As I dream on about the authors of the great books, the mind wanders even further, and I envision Michel Eyquem de Montaigne sitting in front of a personal computer. When I ask him about his presence at our meeting, he starts a long discourse (recalling his essays) about the advantages of word processing for a writer. "Such a tutor will make a pupil digest this new lesson, that the height and value of true virtue consists in the facility, utility, and pleasure of its exercise, so far from difficult, that boys, as well as men, and the innocent as well as the subtle, may make it their own; it is by order and not by force to be acquired." I think he is talking about user-friendly systems and nod my head in assent.

The thread that runs through the lessons of the ages is that we all share a desire to understand our world. In response to that desire, we use whatever tools are available. I am convinced that a great new set of tools has become available. In the tradition of the greats of the past, we will employ them to build new understanding. With no disrespect at all to our predecessors, it should be noted that we are standing at only the beginning of human potential. That is what makes each age so exciting.